Zetetic Cosmogony

Or Conclusive Evidence that the World is not a Rotating Revolving Globe but a Stationary Plane Circle – The Flat Earth Theory Classic

By Thomas Winship

Writing under the Pseudonym of "Rectangle"

Published by Pantianos Classics

ISBN-13: 978-1-78987-298-9

First published in 1899

Contents

Preface to Second Edition

Since the first edition of this work — an unpretentious pamphlet of 48 pages — was published, so much interest in the subject has been manifested, that a second edition is without doubt called for. In fact, long after the first edition was exhausted, letters from various parts of the world, were received, asking for copies, which, to our regret, could not be supplied.

In that pamphlet very much of the evidence we had accumulated from various sources had to be omitted, so as to reduce what otherwise would have been a bulky volume to a short treatise; retaining sufficient evidence to convince the minds of those who would take cognizance of and duly estimate proved facts of nature. Our labours have not been in vain. Many have been enabled to see through the delusions of modern astronomy. Letters from various parts testify that, in some cases, men and women have begun to make use of their brain-power, which had been stunted and dwarfed by acceptation, without the slightest proof, of the unscientific, unreasonable, unnatural, and infidel teachings of men foisted upon a credulous public in the name of "Science." Others again, tell that the writers have thrown to the moles and to the bats the world-wide and almost universally believed hoax that we are living on a whirling sea-earth globe, revolving faster than a cannon-ball travels, rushing through "space" at a rate beyond human power to conceive, and flying with the whole of the so-called solar-system-in another direction twenty times the speed of its rotation.

To the Editors of newspapers, who, whether favourably or unfavourably, reviewed the pamphlet, our thanks are due, and now respectfully tendered.

This edition is sent forth with the assurance of the Divine blessing and the firm conviction that TRUTH IS STRONG AND MUST PREVAIL.

T. W.

12, Castle Buildings,
Durban, Natal,
South Africa,
November, 1899

Introduction

It will be noticed that the style of this volume differs considerably from the first edition. In that edition we divided the book into four parts, viz.: Scientific Assertions, Bible Statements, Natural Proofs, and Application and Conclusion,

The first of these was covered by extracts from well-known astronomical works; the second was filled with Bible quotations, the direct opposite of the astronomical speculations; the third division contained many proofs of the impossibility of the truth of the globe theory; the last division being made up of the logical arguments founded on the first three.

For convenience of reference we have arranged the present edition alphabetically.

In this way any particular branch of the subject found without looking up the index, and something new is found on every page,

Briefly, modern astronomical teaching affirms that world we live on is a globe, which rotates, revolves, and spins away in space at brain-reeling rates of speed; that the sun is a million and a-half times the size of the earth-globe, at nearly a hundred million miles distant from it; that the is about a quarter the size of the earth; that it receives all its light from the sun, and is thus only a reflector, and not a giver, of light; that it attracts the body of the earth and thus causes the tides that the stars are worlds and suns, some of them equal in importance to our own sun himself, and others vastly his superior; that these worlds, inhabited by sentient beings, are without numbers and occupy space boundless in extent and illimitable in duration; the whole of these interlaced bodies being subject to, and supported by, universal gravitation, the foundation and father of the whole fabric.

To fanciful minds and theoretical speculators, the so-called "science" of modern astronomy furnishes a field, unsurpassed in any science for the unrestrained license of the imagination, and the building up of a complicated conjuration of absurdities such as to overawe the simpleton and make him gape with wonder; to deceive even those who truly believe their assumptions to be facts, and to "make men doubt Divine Revelation with as little discrimination as they were formerly called upon to believe."

If the reader will carefully follow and weigh the evidence in the following chapters, he cannot fail to be delivered from the thraldom of popular credulity and led to seek the truth himself.

Current science declares that the earth was once shot off from the sun; a piece of molten rock, which, by universal attraction became larger, "by in draughts from without," as the late R. A. Proctor assures us. This molten mass took 350,000,000 years to cool down for protoplasm to get a footing, which took millions of years "by evolution and selection" to produce a Darwinian ape. Evolution and selection allied to and combined with "the survival of the fittest" again took many millions of years to evolve "primeval" man many ages again elapsing before historical man was produced.

There are four "bodies," according to the late R. A. Proctor, which represent four stages of what we may term astronomical progression, as follow:—

1. The moon was once inhabited, but is now a chaotic mass.

2. The earth is inhabited. It was once like the planet Jupiter. Earlier still it was like the sun, and will become like the moon now is.

3. Jupiter was once like the sun. It is being prepared for inhabitants. When inhabited it will be like the earth. When its race as an inhabited world has been run, it will become like the moon.

4. The sun will become like Jupiter, and another sun will have taken its place. Later it will become like the earth, and will then be inhabited. Later still it will become chaotic like the moon; and so on for countless ages, in fact for ever.

What a grand conception! Yea, rather, what a grand perversion of the reasoning powers, and what stultification of common-sense. What an abuse of precious gifts in order to satisfy a fertile imagination, and supply idle curiosity with something new in the "domain of science."

No one who reads the Bible but can see how these unfounded speculations are diametrically opposed to its plain teaching. The science of the nineteenth century, and the science of the Bible are totally at variance. If the one be true, the other is necessarily false. Which is it: Let the evidence here placed before the reader answer the question. Let honest-minded men and women who read these pages learn the truth for themselves by practical investigation into the facts herein set forth, which we challenge the whole scientific world to successfully dispute.

We court no favour and fear no foe, scientific or otherwise. All we ask is careful attention and practical investigation; we have no fear as to the logical conclusion which will be arrived at.

One - Assumptions

In order to account for natural phenomena in keeping with the assertions of the learned, many hypotheses have to be laid down, and many unfounded assumptions are absolutely necessary to support the unsound fabric of astronomical imagination.

In "Modern Science and Modern Thought," by S. Laing, the following occurs on page 51:—

"What is the material universe composed of Ether, Matter, and Energy. Ether is not actually known to us by any test of which the senses can take cognizance, but it is a sort of *mathematical substance* WHICH WE ARE COMPELLED TO ASSUME IN ORDER TO ACCOUNT for the phenomena of light and heat."

Whatever explanation may be furnished regarding light and heat on this basis, must be discarded as utterly untrustworthy, because the premises are assumed.

Once upon a time it was stated that "the stars were motionless," but as soon as assumption was allowed to talk, the scene was changed, for, as *Science Siftings* informs us (Vol. 6, page 39),

"as soon as it was CONJECTURED that the stars were subject to the law of gravitation, it was inferred that they were not motionless."

Professor Huxley had to resort to assumption to account for the disappearance of ships at sea, although had he known the truth of the matter, or taken the trouble to enquire, his unwarranted assumptions would have been totally unnecessary.

He says:

"We assume the convexity of the water, because we know of no other way to explain the appearance and disappearance of ships at sea."

What learning! What profound wisdom! If we "know of no other way" it is better to admit the fact and wait until we "have found out some other way" to explain the difficulty, if there is any. Knowledge is gained by practical investigation and experience, and has no need of the assistance of assumption to provide an excuse for ignorance. If water could be proved to be convex, there would be no need to assume it to be so. We should have many proofs and abundant evidence of the fact. But the fact that water has been proved to be level, hundreds of times, makes it necessary for those who refuse to believe proved facts which tell against their theory, to resort to assumption to maintain their unreasoning position. And yet this same Professor, in his book "Science and Culture" says:

"the assertion which outstrips evidence is not only a blunder but a crime."

The assertion, therefore, that water is convex against proof furnished many times over that it is level, is not only a blunder, but a crime.

Two - Age of the Earth

This is a subject which has been much speculated upon. I shall quote a few of the more prominent assumptions. Sir Robert Ball, in his "Story of the Heavens," pages 169 and 170, tells us that:

"We cannot pretend to know how many thousands of millions of years ago this epoch was, but *we may be sure* that earlier still *the earth was even hotter,* until at length we seem to see the temperature increase to a red heat, from a red heat we look back to a still earlier age *when the earth was white hot,* back again till we find the surface of our now solid globe was ACTUALLY MOLTEN."

But imagination goes still further than this. In "Our place among Infinities," by R. A. Proctor, pages 9 and 10, we find the following:—

"Let it suffice that we recognise as one of the earliest stages of our earth's history, her condition as a *rotating* MASS OF GLOWING VAPOUR, capturing then as now, but far more actively then than now, masses of matter which approached near enough, and growing by these continual indraughts from without."

How we are to "recognise" that the earth was once a rotating mass of vapour, we are not told. On what evidence the recognition rests, is not stated. Perhaps it is not too much to *assume* that this is like most other assumptions of the astronomical schools, without the slightest vestige of possibility, to say nothing of probability. Sir R. Ball tells us that "we may be sure" that the earth was once "actually molten"; but on what provable data the "surety" of this "actuality" rests we are left to the foggy mazes of imagination to discover. But imagination, assisted by assumption, will account for anything, and so we are told that it "took 350,000,000 years for the earth to cool down from a temperature of 2,000 centigrade to 200." Proctor says that Bischoff has shown this, and so we ought to be sure enough. Were similar ridiculous statements made in relation to any other science than Astronomy or Geology, I believe the general reader would dismiss them at sight. But because they are made in a "domain of science" where the general reader, in most cases, cannot follow, they are allowed to pass as the genuine product of learning and investigation; whereas they are at best but wild and utterly impossible theories. In "Modern Science and Modern Thought," page 44, we are informed that

"It is right, however, to state that ALL MATHEMATICAL CALCULATIONS OF TIME BASED ON THE ASSUMED RATE AT WHICH COSMIC MATTER COOLS INTO SUNS AND PLANETS, AND THESE INTO SOLID AND HABITABLE GLOBES, ARE IN THE HIGHEST DEGREE UNCERTAIN."

Thus, after all the labour to establish a theory, allied with much skill in setting it forth, in its best dress, we are calmly assured that all these tall figures and imaginations are based on premises which are in the highest degree uncertain! If evidence for rejecting these fanciful hypotheses summarily and *in toto* were wanting, surely it is now furnished to satisfaction. Not only are these "mathematical calculations" of assumed premises, "in the highest degree uncertain," but they are to be classed with the tomfooleries of the age, and reckoned among the many and impossible absurdities of the present day.

One of the chief of recent speculations regarding the earth, is that it is a body like the planets, because, it has been shown that the sun and the stars are of the same constituent parts as the earth. Iron, Salt, &c., are said to be elements of the sun's composition, and as the earth contains these and other minerals, it is a globe or planet like the other heavenly bodies which contain the same metals. What is known as

SPECTRUM ANALYSIS

is relied upon as proving this. A prism is placed in position so as to intercept the sun's rays, and the colours seen through this instrument, red, orange, yellow, blue, are said to be the result of the various metals contained in the sun in a state of fusion, emitting their several colours in the combined sunlight, which total light is decomposed into its component colours by the prism.

With the object of testing the conclusions arrived at by the learned relative to spectrum analysis, several experiments were made by the writer. The light of the sun on a clear day, about noon, seen through the prism disclosed the various colours that can be seen through this instrument. On a hazy day before sunset the colours seen were the same but very faint. Light from a lighthouse and a star seen through the prism, showed the colours to be the same, the colour from the light of the star being much less brilliant than that from the lighthouse. Light from a paraffin street lamp gave the same result as light from a star or the sun, only much fainter. Then the electric light was tried. A large street lamp of great power and several others of less power gave the same result as the sun, star, lighthouse, and street lamp, but in various degrees of brilliancy according to the power of the light. Even a candle gave a very faint yellow-blue tinge, so slight that it had to be looked at for some time before anything but blue was apparent.

If, therefore, it be argued that spectrum analysis proves that the sun is made of the same metals as we find in the earth, and that, therefore, the earth is a product of evolution then it is equally clear that the electric light and the glass shade of the lamp which encases it are really composed of iron and various other metals in a state of fusion, constituting indeed, a globe of glowing vapour, and not glass, carbon, &c., at all. It is also as reasonable to conclude that the paraffin lamp and the candle are composed of metals in a state of fusion and that there is in reality no paraffin, no glass, no tallow, and no wick. That is to say, known facts must be thrown aside, common sense stultified, and reason dethroned in order to bolster up the unprovable assumptions of modern science relative to the doctrine of evolution as applied to the earth and the heavenly bodies.

Three - Aeronautics

If the world be a ball, as Sir R. Ball gravely informs us, the aeronaut should be one of his most ardent supporters, as the highest part of the "surface of the globe" would be directly under the car of a balloon, and the sides would fall away or "dip" down in every direction. The universal testimony of aeronauts, however, is entirely against the globular assumption, as the following quotations show. The *London Journal* of 18th July, 1857, says:—

"The chief peculiarity of the view from a balloon at a considerable elevation was the altitude of the horizon, which remained practically on a level with the eye at an elevation of two miles, causing the surface of the earth to appear *concave* instead of convex, and to recede during the rapid ascent, whilst the horizon and the balloon seemed to be stationary."

J. Glaisher, F.R.S., in his work, "Travels in the Air," states: "On looking over the top of the car, the horizon appeared to be on a level with the eye, and taking a grand view of the whole visible area beneath, I was struck with its great regularity; all was dwarfed to one plane; it seemed too flat, too even, apparently artificial." In his accounts of his ascents in the air, M Camilla Flammarion states: "The earth appeared as one immense plane richly decorated with ever-varied colours; hills and valleys are all passed over without being able to distinguish any undulation in the immense plane."

Mr. Elliott, an American aeronaut, says: "I don't know that I ever hinted heretofore that the aeronaut may well be the most sceptical man about the rotundity of the earth. Philosophy forces the truth upon us; but the view of the earth from the elevation of a balloon is that of an immense terrestrial basin, the deeper part of which is directly under one's feet. — *Zetetic Astronomy,* Page 37.

In March, 1897, I met M. Victor Emanuel, and asked him to give me an idea of the shape of the earth as seen from a balloon. He informed me that, instead of the earth declining from the view on either side, and the higher part being under the car, as is popularly supposed, it was the exact opposite; the lowest part, like a huge basin being immediately under the car, and the horizon on all sides rising the level of the eye. This, he admitted, was exactly what should be the appearance of a plane viewed from a balloon.

It is almost needless to say that a globe would present a totally different appearance, the highest part being directly under the car.

Four - Contrasts

If the earth be the globe of popular belief, the same amount of heat and cold, summer and winter, should be experienced at the same latitudes North and South of the Equator. The same number of plants and animals would be found, and the same general conditions exist. That the very opposite is the case, disproves the globular assumption, The *Great Contrasts* between places at the same latitudes North and South of the Equator, is a strong argument against received doctrine of the rotundity of the earth.

From *The Geological Journal for November,* 1893, I extract the following:—

"A Voyage towards the Antarctic Sea," report by Wra. S. Bruce, "On January 12th, 1893, we saw what appeared to be high mountainous land and glaciers stretching from about 64°.10 west to about 65°.30 south, 58° west; this, I believe, may have been the eastern coast of Graham's Land, which has never before been seen. But it would be unwise to be too certain, *for it must have been 60 miles distant.*"

"Meteorology. —Periods of fine calm weather alternate with very severe gales, usually accompanied by fog and snow, the barometer never attained jo inches. The records of air temperature are very remarkable; our lowest temperature was 20°.8 Fahr., our highest 37°.6 Fahr. only a difference of 16°.8 Fahr. in the total range for a period extending slightly over two months. Compare this with our climate; where in a single day and night you may get a variation of more than twice that amount. The average temperatures show a still more remarkable uniformity."

"December averaged 31°.14 Fahr. for one hundred and fifteen readings; January 31°.10 Fahr. for one hundred and ninety-eight readings; February 29°.65 for one hundred and sixteen, a range of less than 1½° Fahr.

This I consider to be very significant, and worthy of special attention by future Antarctic explorers, for may it not indicate a similar uniformity of temperature through the year. Antarctic cold has been much dreaded by some; the four hundred and twenty-nine readings I took during December, January, and February show an I average temperature of only 30°.76 Fahr.; *this being in the very height of summer in latitudes corresponding to the Faroe Islands in the north,* but I believe the temperature in winter will not vary very much from that of summer. This uniformity of temperature partly accounts for the great. accumulation of ice which is formed not on account of the great severity of the winter, but *because there is practically no summer to melt it.*"

"Mr. Seebohm has vividly pictured the onrush of summer in the Arctic; *but how different in the Antarctic.* There, there is eternal winter, and snow never melts. As far north as a man has travelled he has found reindeer and hare basking in the sun, and country brilliant with rich flora; *within the Antarctic circle no plant is to be found.*"

Report by C, W. Donald. M.B., C.M.

"On the passage out, we, on board the *Active,* touched at the beautiful island of Madeira, in October, and two more months landed us in the barren Falkland Islands. Sailing thence on December 11th, we crossed the stormy waters to the east of Cape Horn, and saw our first iceberg on December 18th, On the same day we sighted Clarence Island— one of the South Shetlands. These are called after our own Northern Shetlands, and the part sighted by us lies only some 60 miles nearer the pole. *But what a difference between the two places.* Our own Shetlands bright with ladies' dresses in light summer garments, and carrying tennis racquets and parasols; the South Shetlands, *even in the height of summer,* clad in an almost complete covering of snow, only a steep cliff or bold rock standing out in deep contrast here and there, the only inhabitants being birds or seals; and even the bird life, with the exception of the penguins, is scanty. Sir James Ross, on his third voyage, entered the ice at nearly the same spot, and, fifty years before—all but a week—had sheltered from a westerly gale under the inhospitable shores of Clarence Island. Its highest point stands 4,557 feet above sea level."

The following from "Polar Explorations," read before the Royal Dublin Society, is taken from "Zetetic Astronomy," by "Parallax."

"On the South Georgias, in the same latitude as Yorkshire in the North, Cook did not find a shrub big enough to make a toothpick. Captain Cook describes it as 'savage and horrible.' The wild rocks raised their lofty summits till they were lost in the clouds, and the valleys lay covered with everlasting snow. Not a tree was to be seen; not a shrub even big enough to make a toothpick. Who could have thought than an island of no greater extent than this (Isle of Georgia), situated between the latitude of 54 and 55 degrees, should in the *very height of summer,* be in a manner wholly covered many fathoms deep with frozen snow? The lands which lie to the south are doomed by nature to perpetual frigidness—never to feel the warmth of the sun's rays; whose horrible and savage aspect I have not words to describe. The South Shetlands, occupying a corresponding latitude to their namesakes in the north, present scarcely a vestige of vegetation. Kerguelen, as low as latitude 50 degrees south, boasts 13 species of plants, of which only one, a peculiar kind of cabbage, has been found useful in cases of scurvy; while

Iceland, *15 degrees nearer to the pole in the north, boasts 870 species.* Even marine life is sparse in certain tracts of vast extent, and the sea bird is seldom observed flying over such wastes. The contrasts between the limits of organic life in Arctic and Antarctic zones is very remarkable and significant. Vegetables and land animals are found at nearly 80 degrees in the north; while, from the parallel of 58 degrees in the south, the lichen, and such-like plants only, clothe the rocks, and seabirds and the cetaceous tribes alone are seen upon the desolate beaches." "McLintock describes herds of reindeer—a perfect forest of antlers—moving north in the summer...the eider duck and the brent goose through the air; the unwieldly family of the cetacea through the waters; the Arctic bear upon the ice; the musk ox and reindeer along the land—all wend their way northward at certain seasons Now these indications are absent from the southern zone, as is also the inhabitation of man. The bones of musk oxen, killed by the Esquimaux, were found north of the 79th parallel; while in the south, man is not found above the 56th parallel of latitude."

This is supported by the following from the *Western Christian Advocate,* of 10th February, 1897, copied from *Appleton's Science Monthly.*

"The distinctiveness of the Antarctic climate as compared with the Arctic is found in the relations of both the summer and the winter temperatures. The high summer heat of the north, which in the few months of its existence has the energy to develop that lovely carpeting of grass and flowers which gives to the low-lying lands, *even to the 82nd parallel of latitude,* a charm equal to that of the upland meadows of Switzerland, is in a measure wanting in the south; in its place frequent cold and dreary fogs navigate the atmosphere, and render dreary and desolate a region that extends far into what may be properly designated the habitual zone. The fields of anemones, poppies, saxifrages, and mountain pinks, of dwarf birches and willows, ARE REPLACED BY INTERMINABLE SNOW AND ICE, with only here and there bare patches of rock, to give assurance that something underlies the snow covering. *Man's habitations in the northern hemisphere extend In the 78th parallel of latitude and formerly extended to the 82nd; in the southern hemisphere they find their limit in Fuegia, in* THE FIFTY-FIFTH PARALLEL *fully 350 miles nearer* the equator than where, as in the Shetland Islands, ladies in lawn dresses disport in the game of tennis. And still, 700 miles further from the equator, in Siberia, Nordenskjold found forests of pine riding with trunks 70 to 100 feet in height."

In the "Voyage of a Naturalist," by C. Darwin, pages 210 and 212, we are informed that

"One side of the harbour is formed by a hill about 1,500 feet high, which Captain Fitzroy has called after Sir J. Banks, in commemoration of his disastrous excursion which proved fatal to two men of his party, and nearly so to Dr. Solander. The snowstorm which was the cause of this misfortune, happened in the middle of January, *corresponding to our July in the latitude of Durham.*"

"We were detained here several days by bad weather. The climate is certainly wretched. The summer solstice is now (25th December) passed, yet every day snow fell on the hilts, and in the valleys there was rain accompanied by sleet."

It is utterly impossible to shut one's eyes to the fact that these evidences furnish indisputable proof that the figure of the earth cannot be globular. If it were of that shape the same conditions would be found at equal latitudes north and which we have seen is not the

Five - Contradictions

The grave contradictions that exit among the recognised teachers of astronomical science, ought to cause a thinking man to pause before accepting a theory about which no two of its exponents may be found to agree.

Sir Isaac Newton, in his "Principia," resuscitated the fundamental proposition of Pythagoras thus— "The sun is the centre of the solar system and immovable." Since then Professor Herschel discovered that the sun was "*not immovable.*"

In regard to the atmosphere of the planet Mars, the same contradiction is manifest. In the *Christian Millton* (San Jose) of 9th August, 1894, we find that

"Mr. Norman Lockyer has been telling an interviewer that Mars is like us in many respects. IT HAS AN ATMOSPHERE LIKE OURS."

The *Standard* of 18th August, 1894, says:—

"Professor Campbell, of the Lick Observatory, announces that he has demonstrated that MARS presents NO EVIDENCE OF HAVING AN ATMOSPHERE."

Then Mr. J. Gillespie, in his "Triumph of Philosophy," page 89, comes to the rescue and says:

"As to the planets being inhabited, if we take refraction into account, we shall find that *there is not such a thing as atmosphere near them;* for instance, in an eclipse of the moon, especially at her apogee, the earth is brought to a mere point by refraction, caused by the air of the earth, and were the moon a little further away from this point, would be brought to nothingness; that is although the earth were exactly in a straight line between the sun and moon, the earth would not even show a spot on the moon's disc...Now by this same rule, if either Mercury or Venus had any atmosphere, they could *never* be seen crossing the sun's disc. I think this is satisfactory proof that THEY HAVE NO ATMOSPHERE, *and cannot, therefore, be inhabited.*"

After all this delightful uncertainty, a writer in *Knowledge* of February, 1895, says:

"The interesting chapter on solar theories is well fitted to serve as a lesson in *modesty,* so diverse and conflicting are the various hypotheses, *so difficult to harmonise,* are the observed facts."

When we come to consider the atmosphere that concerns us most, the same contradictions are evident. Sir David Brewster, in his "More Worlds than One," tells us that the atmosphere of the earth extends for about 45 miles. In *Science Siftings* of 18th March, 1893, the following occurs:

"We may *infer* that a few hundred miles embrace all the gaseous envelope of the globe."

And in "Elementary Physiography," page 293, we are told that

"*The height of the atmosphere is not known* with any certainty. *There is probably no fixed limit* to the atmosphere."

It is a fair inference from these contradictory statements that present day scientists (so-called) *do not know anything* about the height of the earth's atmosphere.

Many men of thought and learning have scouted the ideas imposed upon us by Sir Isaac Newton, of which the following is a sample:—

"The repetition of a blunder is impertinent and ridiculous. To liberate oneself from an error is difficult, sometimes indeed impossible for even the strongest and most gifted minds. But to take up the error of another, and persist in it with stiff-necked obstinacy is a proof of poor qualities. The obstinacy of a man of originality when he errs may make us angry, but the stupidity of the copyist irritates and renders us miserable. And if in our strife with (Sir Isaac) Newton, we have sometimes passed the bounds of moderation, the whole blame is to be laid upon the school of which Newton was the head, whose incompetence is proportional to its arrogance, whose laziness is proportional to its self-sufficiency, and whose virulence and love of persecution hold each other in perfect equilibrium." Through the whole of Newton's experiments there runs display of pedantic accuracy, but how the matter really stands, with Newton's gift of observation, and with his experimental aptitudes, every man possessing eyes and senses may make himself aware. It may be boldly asked, where can the man be found, possessing the extraordinary gifts of Newton, who would suffer himself to be deluded by such a *hocus pocus* if he had not in the first instance wilfully deceived himself? Only those who know the strength of self-deception, and the extent to which it sometimes trenches on dishonesty, are in a condition to explain the conduct of Newton and of Newton's school. To support his unnatural theory, Newton heaps fiction upon fiction, seeking to dazzle when he could not convince."—Goethe. - Proceedings of the Royal Institute of Great Britain. Vol. ix., part iii., p. 353-5.

Dr. W. Friend says

"It has, over and over again, been the hope and expectation of intelligent and unprejudiced men that some less extravagant and more intelligible system would, sooner or later, be found as a substitute for the mathematical romance with which Newton has favoured the world. This name has been the sanction for a device, which, the more it is examined, excites the more astonishment at its adoption by men of research and observation.

Then, again, Kepler's laws, said to be so well established and so absolutely necessary to the truth of the Newtonian hypothesis, when weighed in the balance by competent judges, are contradicted and set aside by a stroke of the pen. Professor W. B. Carpenter, in the *Modern Review* for October, 1880, says:

"It was not until twelve years after the publication of his first two laws, that Kepler was able to announce the discovery of the *third.* This, again, was the outcome of a long series of GUESSES, and what was remarkable as to the error of the idea which suggested the second law to his mind, was still more remarkable as to the third; for not only, in his search for the 'harmony' of which he felt assured, did be proceed on the erroneous notion of a whirling force emanating from the Sun, which decreases with increase of distance, but he took as his guide ANOTHER ASSUMPTION no less erroneous, viz., that the *masses* of the Planets increase with their distances from the Sun. In order to make this last fit with the facts he was I ASSUME a relation of their respective *densities,* which we *now know to be* UTTERLY UNTRUE; for, as he himself says, 'unless we ASSUME this proportion of the densities, the law the periodic times will not answer.' Thus, says his biographer, 'three out of the four *suppositions* made by Kepler to explain the beautiful law he had detected are now INDISPUTABLY KNOWN TO BE FALSE, what he considered to be the *proof* of it being only A MODE OF FALSE REASONING by which 'any *required result* might be deduced from any given principles."

Newton's theory and Kepler's laws are the chief foundation stones of modern astronomy, and when these are shaken, the whole fabric reels and staggers like a drunken man; until, sooner or later it will find a grave in the oblivion that it so well merits.

The *Daily Chronicle* of 8th April, 1891, says;

"It may be a surprise to find that we are still imperfectly acquainted with the figure of the Earth."

The *Ceylon Independent,* of 23rd December, 1893, has the following:—

"This question seems to be still agitating the Austrian Government, and more than one Austrian man-of-war that has called here lately has had an officer on board whose special commission was to make observations for the purpose of ascertaining the attraction of the earth in order thereby to arrive at the exact shape of the globe. An officer thus employed is on the Austrian steamer *Fasana,* who, since the vessel's arrival, has spent a good deal of time at the National Bank, where a room was allotted him for the purpose of adjusting his instruments. An officer engaged on similar duty was on the *Kaiserin Elizabeth* the other day."

Von Gumpach, in his work "Figure of the Earth," tells us how the men of science made the world a globe.

"The earth of the Newtonian theory, is the mere creation of the fancy. Its shape has been determined, partly of imaginary and partly of positively erroneous elements; and results of subsequent experiments and measurements have, by means of purely mathematical factors and tentative formulas been adapted to its PRE-SUPPOSED FIGURE."

Mr. J. Gillespie, who believes that the earth is a globe suspended in space, with no revolution round the sun, says, in his "Triumph of Philosophy," p. 6.

"I can challenge any astronomer in Great Britain on any point in theoretical astronomy, and prove that *the present theory is a regular burlesque,* A HOAX and A SWINDLE. If it is a sin to tell a lie, what must be the doom of men who teach generation after generation one of the most glaring and degraded falsehoods ever laid before mankind."

Dr. Lardner, in his Museum of Science," says

"All the diurnal changes of appearances, presented by the firmament, the risings and settings of the sun and their varying appearance of different latitudes, admit to being explained with equal precision and completeness, either by supposing the universe to revolve daily round the earth, or earth to revolve daily on its axis."

Then as to the velocity of light (if light travels at all), the same glorious mixture and uncertainty again present themselves. Guillemin ("The Heavens") conjectures that light travels at the rate of 192,000 miles a second. M. Leon Foucault guesses 184,000 miles; Sir R. Ball 180,000 miles; the Editor of *Science Siftings* assumes (first time) 186,000 miles, second time 196,000 miles. This is all contradicted by a writer in the *English Mechanic* of 27th July, 1894, who says:

"I BELIEVE NO ONE NOW HOLDS THE VIEW THAT LIGHT ACTUALLY MOVES."

Most people think that there is only one school of Astronomy in vogue, whereas there are at least four, all at loggerheads with each other, (1) The Ptolemaists, represented by J, Gillespie, of Dumfries, who suppose the "earth" globe a centre for the revolution of the sun, moon, and stars; (2) The Koreshans of America, who suppose the "earth" a hollow globe for us to live inside; (3) The Newtonian Copernicans, who suppose the sun a centre, keeping the planets whirling in orbits by gravity; and (4) the Cartesian Copernicans, who suppose the planets to whirl round the sun, without the necessity of gravity. Sir R. Phillips heading up this school.

Astronomy will sometimes summon Geology to its aid when difficult problems are awaiting solution, but astronomers generally claim that when the two sciences disagree, astronomy is the *safest* ASSUMPTION. S. Laing, however, as "Modern Science and Modern Thought" claims superiority for Geology. On pages 48 and 49, he says:

"The conclusions of Geology, at any rate up to the Silurian period are *approximate facts* and NOT THEORIES while the astronomical conclusions are THEORIES, *based on data so uncertain,* that while in some cases they give results *incredibly short,* like that of 15,000,000 years for the whole past process of the formation of the solar system, in others they give results almost *incredibly long,* as in that which supposes the Moon to have been thrown off when the earth was rotating in three hours...the *safest course,* in the present state of our knowledge seems to be to ASSUME THAT GEOLOGY REALLY PROVES the duration of the present order of things to have been somewhere over 100,000,000 years."

Thus one fable (falsely called science) exposes another fable of about the same value. "The *safest course* in the present state" of the utter ignorance of "science" as to the matters here in dispute, is certainly to reject both these delusions, and seek the truth for ourselves.

Geological blunders have been many and frequent, but they are seldom allowed to reach the eyes or ears of those who are duped into believing all this imposing "science" teaches. *The Daily Chronicle* of 14th January, 1893, speaks pretty plain, and proves the truth of the above remarks. The paper says:

"A GEOLOGICAL BLUNDER."

"There is in *Nature* an article by a French writer on Sir Archibald Geikie, Director-General of the Geological Survey, which is just now causing a good deal of talk amongst English men of science. Of course, nobody is surprised at the fulsomeness of M. de Lapparent's eulogy. As *Nature* seems to exist for pushing the great official scientific syndicate of Huxley, Hooker, Geikie and Co., Limited—very strictly limited which may be said to "run" science in England, M. de Lapparent would probably not have been *permitted* to write anything about a member of it unless it was fulsome. What has really amazed people is the audacity with which a famous historic bungle on the part of the Geological Survey is glossed over, and the Director-General not only credited with the work of those who exposed and corrected it, to his utter discomfiture, but actually covered with laurels for thus winning one of the most glorious scientific conquests of the century. The whole thing is delightfully characteristic of State-endowed science in England. If you are one of the official syndicate who "run" it, you may blunder with impunity and make your country ridiculous at the taxpayers' expense. Scientific men who can correct you shrink from the task. They know that the syndicate can *boycott* them, and by *intrigue* keep them out of every honour and profit, and that the syndicate's satellites can write and *shout down* everywhere independent non-official critics. They also know that if, perchance, some particular intrepid person does succeed in exposing one of this syndicate, they can always, by the same means—after the public has forgotten the incident suppress him, and boldly appropriate to themselves the credit of his work."

"The geological secret of the Highlands, with the unlocking of which Sir Archibald Geikie is now credited, was really made a puzzle for more than half a century by the blundering of the Geographical Survey and Director-General Sir Roderick Murchison—and famous courtier and "society" geologist of the last generation. In the Highlands he saw gneisses and ordinary crystalline schists resting on

Silurian strata, and he foolishly held the sequence to be quite normal, The schists, he would have it, were not archaic formations, but only metamorphosed Silurian deposits. He also held that primitive gneiss was not part of the molten *crust of the globe*, but only sediments of sand and mud altered by intense pressure and heat. Murchison, not to put too fine a point on it, "bounced" everybody into accepting this absurd theory, and the whole forces of the Geological Survey, with its official and social influence, together with the unscrupulous power of the official syndicate which then, as now *jobbed* science wherever it had a State endowment, were spent in perpetuating the blunder and blasting the scientific reputation of whoever scoffed at it. But in the Natural History School of Aberdeen University it *was* scoffed at. The late Dr. Nicol, Professor of Natural History in Aberdeen, proved that Murchison and the Survey were *wholly wrong,* his proof being as complete as the existing state of science allowed. When he died, Dr. Alleyne Nicholson took the same side, and for years, in relation to this grand problem, it was Aberdeen University against the world...In shouting the last word no voice has far louder than Sir Archibald Geikie's. It is therefore diverting find his official biographer stating in *Nature* that all the time he was wrestling *in foro conseientiae* with doubts as to the soundness of the official position, and that finally "his love of truth" prompted him to order a re-survey of the whole Highland region. In plain English, the taxpayer, having had to pay for Murchison's bungling survey, was, because of his successor's "love of truth," to enjoy the luxury of paying over again to correct it.

The real truth, however, is this:—When it was supposed that the Aberdonians were finally crushed, there arose in England a young geologist called Lapworth, who had the courage to revise the whole controversy and take sides with the Aberdeen School. As he developed an extraordinary genius for stratigraphy he not only broke to pieces the official work of the Geographic Survey in the Highlands, but by revealing the true secret of the structure of that perplexing region, he played havoc with the Murchisons and the Geikies and all their satellites, convicting them of *bungling* and covering them with ridicule...

Nature, in fact, in these parts had suffered from a much more powerful emetic than Murchison imagined, and when bits of the primitive artist of the glome were thrown up and pushed on the top of more recent deposits, Murchison *jumped lo the conclusion* that they were of later date than what they lay on. It was a *terrible blunder,* as the Aberdeen men persistently held, and we do not wonder that Sir Archibald Geikie, who rose to place and power by defending it, is anxious to have his connection with it *veiled* by a friendly hand. But it is rather outrageous for the friendly hand to give him the credit of conceding the very *error* which he *defended to* the *last gasp*, and deprive Professor Lapworth of the honour of having banished it from science. One of the most diverting things, however, in the Article in *Nature* is that Sir Archibald Geikie is belauded because, when frightened by the stir Professor Lapworth's paper made in 1883, he was fain to send his surveyors to go over the Highlands again—he, as their official chief, ordered them "to divest themselves of any *prepossession in favour of published views,* and to map out the actual facts." Old Colin Campbell, when he objected to the institution of the Victoria Cross, said it was as absurd to decorate a soldier for being brave as a woman for being virtuous. He did not foresee a still greater absurdity—that of eulogising a man of science because he instructed his assistants to tell the truth when conducting an investigation into his own blunders." (Italics ours). – From the *Daily Chronicle,* Saturday, Jan. 14th, 1893.

And in a further issue the same paper says:

"Sir Archibald Geikie, Director-General of the Geological Survey, has at last taken notice—in *Nature,* we need hardly say —of our article condemning the attempt to give the Survey all the credit of some of the most remarkable discoveries of the age which have really been made by men unaided by the State, and toiling for daily bread as teachers of science. We bad heard something that caused us to expose this scandal. The fact is the official ring of State-endowed science, not content with *jobbing* the Royal Society and its distinctions, as their critics have been showing in the *Times,* are meditating a raid on the taxpayer. They want more money, and as a preliminary step their official organ *Nature* of course begins to "boom" their work and reputations. This is a good old game. The only novelty in the situation is that a daily newspaper, for the first time in history, ventured to show it up. We do not desire to be harsh to the illustrious scientists who edit *Nature.* It is the duty of all official organs to make big men out of small material. But when they begin to do this by coolly confiscating the achievements of private and independent workers for one of the managing partners of the great firm of Huxley, Geikie, Dyer & Co., limited, we thought it time to protest...The letters that have been appearing in the Times make some funny revelations about the way the Royal Society is "worked." Sir Archibald Geikie's defence suggests that if the *Times* only followed up the game it scented it would show its readers plenty of sport. We ourselves would make no objection to a vote of money in aid of researches into the "frank" and "practical" manner in which, *and the terms on which,* the official gang of science frequently "acknowledge" the achievements of young outsiders."— *Daily Chronicle,* Feb. 2, 1893.

Modern Astronomy has been set down as "the most exact of all the sciences," and geology said to be little less than infallible. The reader may form his own conclusion from the above extracts.

Six - Circumnavigation

is said to be one of the best proofs that the earth is a globe.

It is often asserted—generally by those who have not the remotest idea of the subject—that ships have sailed round the world on one course, East or West' and come back to the place where they started from. It will be a surprise to such to be informed that this wonderful feat of navigation has never yet been accomplished; that it is most unlikely that It will ever become a fact; and that it would take several of the proverbial "small fortunes" to successfully carry it out.

Some people think it is quite an easy matter to start from, say Liverpool, and steer west and come back to the starting point. Suppose we attempt such a journey. After crossing the Atlantic we must leave the ship and traverse the American continent As there are no roads running due west, we should have to take the sun's bearing almost hourly to keep us on the true course; sometimes having to cross private property, travel through cultivated lands, and in acme cases to go through other people's houses to preserve a westerly course. Suppose we arrived at the other side and then took ship across the Pacific, we should again have to travel across a continent — thousands of miles — to get back to the North Sea, and then across it and England we might arrive at Liverpool. If anyone thinks this possible he ought to try it.

If the reader will scan the surface of a school globe, he will at once see that if such a thing should ever be attempted, no reasonable hope of success could be entertained, unless the attempt were made in the extreme south. Suppose a ship to start from Cape Point, latitude 34 south, and steer east. The first land encountered would be Australia, ships would then have to go south to clear the land and so could not return to her starting point on an easterly course, but would have to take many courses to return there.

Let the ship start from Cape Horn, in latitude 56 south and steer west. She would soon encounter islands and would have to alter her course to north or south to clear them, so could not get back to Cape Horn on a westerly course. The same would apply on an easterly course.

It is evident, therefore, that the earth can only be circumnavigated on one course in the extreme south. There, the dangers of icebergs, of magnitudes never met with in the north, and darkness during a great part of the year, would render such an expedition costly, dangerous, and of long duration.

Say a vessel starts on an easterly or westerly course in latitude 65 south. She could only sail during the very finest of summer weather, and would have to come north I during the winter. Returning to her last point, she could again start on the course round the world, and continue so long as the fine weather lasted, repeating the process of going north during the dark and winter months. That this would occupy a long time, and cost a deal of money, is plain enough to anyone willing to be convinced. *For these reasons I am of the opinion that no ship will ever sail round the world on one course and come back to her starting point.* And yet to some the world is a globe! One of the greatest feats of navigation will tell you that it has been done scores of times, and proves and seamanship that man could undertake, and which has never yet been attempted, is spoken of as though it were a matter of almost daily occurrence! And who but the astronomers are responsible for such-like fallacies in school books and astronomical works? Who but those famed for "learned ignorance" are answerable for the foolish arrogance and stupid credulity of the masses on this subject? Can there be any truth in a science which is founded on conjecture and supported by so-called facts as proof of its correctness, which facts have never existed outside the brains of their inventors?

If it were said that a vessel could sail round the world, allowing for deviations for land, ice, and other obstacles in the way of her making one course; so that by making many and various courses she could at length return to her starting point, I would have no quarrel with the propounders of "circumnavigation." But if the general statements on the point were reduced and brought within the compass of fact, in language such as the above, the supposed proof of the world's rotundity would be annihilated. In Evers' "Navigation" it is stated that a vessel may leave a port, sail round the earth, and come back to her starting point *on one course.* This, I have no hesitation in stating, is *absolutely false.* If otherwise, I should be glad to be informed of the name of the port.

The learned are beginning to see through the fallacy of the circumnavigation proof of the world's rotundity, as the following from "Elementary Physiography," by Professor Richard A, Gregory, F.R.A.S., clearly shows:

"The earth has been circumnavigated a great many times, and it is a common occurrence for a ship to leave England, and by steering westward all the voyage to arrive in England again without retracing an inch of her way. Similarly, we can journey round the globe, sometimes traveling on land, and sometimes on the sea, but eventually returning to the starting point without at all turning back on our course. This would appear to be a certain proof that the earth's surface is curved, nevertheless it has been pointed out that circumnavigation would be possible if the earth had a flat surface, with the north magnetic pole at its centre. A compass needle would THEN *always point* to the centre of the surface, and so a ship might sail due east and west, as indicated by the compass, and eventually return to the same point by describing a circle."

D. Wilson-Barker, R.N.R., F.R.S.E., remarks, in his work on "Navigation":

"The fact that the earth has been sailed round, is not sufficient proof as to its exact shape."

After these "authoritative" statements, we may hope that this so-called proof of the globular shape of the earth will soon be expunged from the text books.

Seven - Curvature

In "Chambers' Mathematical Tables" the curvature of the globe is given as 7.935 inches to the mile, varying inversely as the square of the distance. If it be required to ascertain the curvature on a globe of 25,000 statute miles equatorial circumference, square the distance and multiply by 7.935 inches. The result is the curvature. Thus, in six miles there is a dip of nearly 24 feet; in 30 miles, nearly 600 feet; and so on.

In "Mensuration," by T. Baker, C.E., the correction for curvature is said to be 7.962 inches to the mile. These two equations so nearly agree, and amount to just about what the correction would be on a globe of the size the earth is said to be, that they may be taken as correct. If, therefore, the world we live on is a globe, it is a simple matter to find out how far any object at a given height can be seen.

In September, 1898, I received a letter from Australia, in which the writer says:

"In the year 1872 I was on board the ship "Thomas Wood," Capt. Gibson, from China to London. Owing to making a long passage, we ran short of provisions, and so short after rounding the Cape that the Captain spoke of putting into St. Helena for a supply. It was then my hobby to get the first glimpse of land, and in order to do this I would go up to the topgallant yard and make a survey, just as the sun would be rising. The island was clearly in view, well on the starboard bow. I reported this to Capt. Gibson. He disbelieved me, saying it was impossible, as we were *75 miles distant.* He, however, offered me paper and pencil to sketch the land I saw. This I did. He then said, 'you are right,' and shaped his course accordingly. I had never seen the Island before, and could not have described the shape of it had I not seen it."

St. Helena is a high volcanic island, and if my informant had seen the top only, there would have to be an allowance made for the height of the land, but as he sketched *the island,* he must have seen the whole of it, which should have been 3,650 feet below the line of sight, if the world be a globe (deducting 100 feet for the height of the yard he viewed it from).

In "Chambers' Information for the People," section t Physical Geography, page 513, the following occurs:

"In North America, the basin or drainage of the Mississippi is estimated at 1,300,000 square miles, and that of the St. Lawrence at 600,000; while northward of the 50th parallel, extends an inhospitable *flat* of perhaps greater dimensions...Next in order of importance is that section of Europe extending from the German Sea, through Prussia, Poland, and Russia, towards the Ural Mountains, presenting indifferently tracts of heath, sand and open pasture, and regarded by geographers as ONE VAST PLANE. So flat is the general profile of the region, that it has been remarked, IT IS POSSIBLE TO DRAW A LINE FROM LONDON TO MOSCOW, WHICH WOULD NOT PERCEPTIBLY VARY FROM A DEAD LEVEL."

The foregoing is a London-to-Moscow proof that the surface of the world is not globular. On a globe, no matter how powerful the glass, only a certain distance could be seen, as the roundness of the globe would prevent a glass from seeing round it, and its thickness would equally prevent one seeing through it. But in fine weather objects at distances out of all proportion to what the curvature would allow, are visible with the assistance of a good glass. The following from the "Voyage of a Naturalist," by C. Darwin, page 166, illustrates this point:

"The guanaco, or wild llama.—Mr. Stokes told me that he one day saw, through a glass, a herd of these animals which evidently Sad been frightened, and were running away at full speed, although *their distance was so great that he could not distinguish them* with the naked eye."

From the "Atlas of Physical Geography," by the Rev. T. Milner, M. A., I extract the following:

"Vast areas exhibit a *perfectly dead level,* scarcely a rise existing through 1,500 miles from the Carpathians to the Urals, South of the Baltic *the country is so flat* that a prevailing north wind will drive the waters of the Stattiner Haf into the mouth of the Oder, and give the river a backward flow 30 or 40 miles."

"The plains of Venezuela and New Granada, in South America, chiefly on the left of the Orinoco, are termed llanos, or level fields. Often in the space of 270 square miles THE SURP'ACE DOES NOT VARY A SINGLE FOOT."

"The Amazon only falls 12 feet in the last 700 miles of its course; the La Plata has only a descent of one thirty-third of an inch a mile."

These extracts clearly prove that the surface of the earth is a level surface, and that, therefore, the world is not a globe. And when we come to consider the surface of the world *under the sea,* we shall find the same uniformity of evidence again the popular view. In "Nature and Man," by Professor W. B. Carpenter, article "The Deep Sea and its Contents," pages 320 and 321, the writer says:

"Nothing seems to have struck the "Challenger" surveyors more than the extraordinary FLATNESS (except in the neighbourhood of land) of that depressed portion of the earth's crust which forms the FLOOR OF THE GREAT OCEANIC AREA...If the bottom of mid-ocean were laid dry, *an observer standing on any spot of it would find himself surrounded* BY A PLAIN, only comparable to that of the North American prairies or the South American pampas...The form of the depressed area which lodges the water of the deep ocean is rather, indeed, to be likened to that of a FLAT WAITER or TEA TRAY, surrounded by an elevated and deeply-sloping rim, than to that of the basin with which it is commonly compared."

This remarkable writer tells of thousands of miles, in the Atlantic, the Pacific, and the great Southern Ocean beds being a plane surface, and from his remarks it is clear that A FLAT SURFACE IS THE GENERAL CONTOUR OF THE BED OF THE GREAT OCEANS FOR TENS OF THOUSANDS OF SQUARE MILES.

Eight - Canals

If the earth be the globe of popular belief, it is very evident that in cutting a canal, an allowance must be made for the curvature of the globe, which allowance would correspond to the square of the distance multiplied by eight inches, nearly. From the *Age,* of 5th August 1893, I extract the following:

"The German Emperor performed the ceremony of opening the Gates of the Baltic and North Sea Canal, in the spring of 1891. The canal starts at Holtenau, on the south side of Kiel Bay, and Joins the Elbe 15 miles above its mouth, It is 61 miles long, 200 feet wide at the surface and 85 feet at bottom, the depth being 28 feet. No locks are required, as *the surface of the two seas is level."*

Let those who believe it is the practice for surveyors to allowance for "curvature" ponder over the following from the Manchester Ship Canal Company. — (*Earth Review,* October, 1893).

"It is customary in Railway and Canal constructions for all levels to be referred to a datum which is nominally horizontal, and is so shown on all sections. *It is not the practice in laying out Public Works to make allowance for the curvature of the earth."* — Manchester Ship Canal Co., Engineer's Office, 19th February, 1892."

A surveyor, Mr. T. Westwood, writes to the *Earth Review,* for January, 1896, as follows:

"In levelling, I work from Ordnance marks, or canal levels, to get the height above sea level...I work sometimes from what is known as the Wolverhampton level, this is said to be 473.19 feet above sea level; sometimes I work from the Birmingham level, this is said to be 453.04 feet above sea level. Sometimes I work from the Walsall level, this is said to be 407.89 feet above sea level. The puzzle to me used to be, that, though each extends several miles, each level was and is treated throughout its whole length as the same level from end to end; not the least allowance being made for curvature, although if the earth were a globe, 112 feet ought to be allowed...One of the civil engineers in this district, after some amount of argument on each side as to the reason why no allowance for curvature was made, said he did not believe anybody would know the shape of the earth in this life."

I think most will grant that a practical man is capable of forming a judgment, in all cases of more value than the merely theoretical calculator. Here, then, we have the evidence of practical men to the effect that no allowance for curvature is made in cutting canals, a clear proof that we are not living on a huge ball, but on a surface, the general contour of which is level, as the datum line from which surveys are made IS ALWAYS A HORIZONTAL LINE.

Nine - Disappearances of Ships at Sea

J. W. Draper, in his "Conflict between Religion and Science," page 160, says:

"The circular visible horizon and its dip at sea, the gradual appearance and disappearance of ships in the offing, cannot fail to incline intelligent sailors to a belief in the globular form of the earth."

The "circular visible horizon" amounts to nothing, because if we take our stand in a large square of, say, 20 miles, the visible horizon will be circular, any point in the distance being the edge of the circle of vision. If we measure off a square of 100 miles or so, the vision will be bounded by a circle, the limit of sight. So the "circular visible horizon" may at once be dismissed. But "its dip at sea" is just what has never been seen. It is the very thing that requires to be seen to establish the globular theory; it is the very thing that never has been seen. Wherever we look at sea, the water extends in one straight line, as far as the eye can reach. A flat surface is always seen, and ships are seen at distances altogether out of proportion to the allowance to be made for convexity, if the surface were a convex one.

When a ship or any other object recedes from the observer on a level surface the highest part is always last by reason of perspective. So that the masts and sails of; a receding vessel on a flat surface should be seen long after the hull has become invisible to the naked eye. Besides this law of perspective, the hull of a vessel is generally of a dark colour, and often at a very short distance disappears to the naked eye, because it has lost its individuality in the mass of surrounding water, both hull and water being nearly of the same colour. It appears to have mingled with the water, and is thus lost to sight. The hull has no background whatever, but the masts and sails have a splendid background against the sky, and stand out to advantage, and are, for this reason also, seen long after the hull has vanished. But that the hull has not "gone down behind a hill of water"—that it is not because of the globular surface of the water that it is invisible—has been proved by the writer many times.

At Capetown, sometime ago, I made special experiments with a view to arrive at the truth of the matter. On one occasion I watched the schooner *Lilla,* of Capetown, sail away north, bound to Saldanha Bay. Instead of gradually going down the hill of water—the observer always being on the highest part—she appeared to ascend an inclined plane, until she reached the level of my eye—perhaps 100 feet above sea level—and then gradually diminished in size. Soon her hull disappeared—it was painted black—and her masts and sails became smaller and smaller every minute. I then applied a binocular to the eye, and saw her hull plainly enough. It remained in sight until the individuality of the vessel's parts were lost in the distance.

The iron barque *La Querida,* of Liverpool, sailed out of Table Bay bound to Australia. I watched her until the hull had completely disappeared; but on applying the glass saw it as clearly as possible, and this when the vessel was at least 10 miles away. So that the "hill of water" in both these instances was imaginary only.

In May, 1895, I was a passenger on board the U. S. S. Goth. In Algoa Bay I gave a brief lecture on the subject of this work, and had much discussion with some of the passengers; one affirming he could believe all I said, with the exception of the way I accounted for the disappearance of ships at sea. I replied that we would likely see one of the ships, and then it could be tested. Next day I observed a vessel about ten miles away, but though the masts and sails were pretty clear, the hull was not to be seen. Applying the glass I saw the hull as plain as any other part of the ship, I called the gentleman with whom I had the previous day's conversation and showed him the vessel. I asked him to look at the ship for some time so as to be quite sure whether the hull was visible or not. After looking a minute or so he was quite certain that the hull could not be seen I asked him why it was invisible. "Because," said he, "it is hidden behind a hill of water, the surface of the ocean being: convex." I asked him if he believed my glass could see through a "hill of water," and gave him the astronomer's curvature for the distance—which he admitted to be lo miles—as 10 by 10 by 8 inches = 66 feet, less 20 feet for height of eye and 10 feet for height of the other vessel's hull, = 36 feet the hull should have been below the water. He replied that the glass could not, of course, see through a hill of water, and applied it to his eye. Great was his astonishment on seeing the hull, but equally ready was his confession that the theory of the earth's rotundity founded on the disappearance of ships at sea was false.

On a steamer in March, 1897, when near St. Helena my attention was called to a large vessel, just before sunset. With the naked eye the masts and sails were visible enough, but nothing of the hull could be seen. On applying the glass, there appeared to be no difference, and I was for some time lost in wonder. But as the sun got lower in the heavens, I noticed that the vessel's hull was overshadowed by banks of black clouds low down on the water and thus could not be seen. The hull was enveloped in dense blackness and was lost to the eye. But as soon as the sun was low enough to counteract this effect, I saw the hull quite plain with the glass, when only the sails were visible to the naked eye.

Between Tenerife and Southampton we sighted a large four-masted steamer astern of us. The hull was also plainly to be seen — the vessel appeared to be in ballast. Our ship's officers said she was 12 miles away, and I think the distance was not less. For two whole days she was visible to us astern; sometimes the hull being quite plain, at other times being invisible; thus proving that the state of the atmosphere has more to do with the matter than globularity, if it existed, could have. *According to the globe theory, an object plainly visible to the naked eye and seen by scores of people should have been 96 feet below the horizon,* allowing both vessels to be the same height above the water, which was as near as possible correct, as our ship had scarcely any cargo on board and presented a high side out of the water.

ANOTHER WITNESS

"To the Editor of the *Earth Review,*

Sir,—In August last I, with several other friends, being in Oban for a holiday, took a trip for a day in a small yacht on Loch Lorne, and being a glorious sunshiny day and so calm that not a ripple was seen, and being becalmed for an hour about mid-day we observed a good many sights of various kinds. Amongst other things that we saw was a yacht, which the captain told us was 12 miles distant. We saw all the masts and part of the hull, and to get a better view of her we took our binocular opera glass (a good one). Now, sir, wouldn't it require a funny curvature table either with or without the odd fractions to explain how we saw the hull of that vessel twelve miles off? According to a table furnished by the present Astronomer Royal recently, it ought to have been 66 feet below the line of sight; but the "table" that we saw it from was the side of our yacht, and we concluded the sea was level.

Yours respectfully,

Siddal, Halifax. JOHN SMITH.

The following is from "100 Proofs that the Earth is not a Globe":

"If we take a trip down the Chesapeake Bay, in the daytime, we may see for ourselves the utter fallacy of the idea that when a vessel appears "hull down," as it is called, it is because the hull is "behind the water": for, vessels have been seen, and may often be seen again, presenting the appearance spoken of, and away—far away—beyond those vessels, and, at the same moment, the level shore line, with its accompanying complement of tall trees, towering up, in perspective, over the heads of the 'hull-down' ships!"

The following is from *Chambers' Journal* of February, 1895, page 32:

"A good many years ago a Pilot in the Mauritius reported that he had seen a vessel which turned out to be 200 miles off. This incident caused a good deal of discussion in nautical circles at the time, and strange to say, a seemingly well authenticated case of the same kind occurred afterwards at Aden. A Pilot there announced that he had seen from the heights the Bombay steamer then nearly due. He stated precisely the direction in which he saw her, and added that her head was not then turned towards the port...Two days afterwards the missing steamer entered the Port, and it was found often on enquiries that at the time mentioned by the Pilot she was exactly in the direction and position indicated by him, but ABOUT TWO HUNDRED MILES AWAY."

Under exceptional conditions of the atmosphere, therefore, enormous distances can be penetrated by the unaided eye, and with a good telescope, objects at distances totally out of proportion to the globular theory, can be seen. Take the case of the above steamer. If the globe theory be correct this vessel would have been FOUR MILES BELOW THE LINE OF SIGHT, *allowing one mile for height of observer,* and thus even when aided by the most powerful telescope ever invented, could not have been seen. Once more, it dawns on the thinking man, that the world is not the globe of popular credulity, but an extended motionless plane.

Ten - Distances

If the world be a globe, the distances which are sailed by ships "sailing round the globe" would answer to the theory, and measurements as made by such ships would always answer to the theoretical distances of the astronomer. That such is not the case, as I shall presently show, disproves the theory. First, let us enquire how distances are obtained, say in sailing on an easterly or westerly course. In obtaining the longitude by dead reckoning, an allowance for the supposed convergence (or shorter longitude) according to the latitude would have to be made, when the result obtained should not vary much from longitude obtained by observation. When currents have to be reckoned with, the allowance for their known velocity in any direction would bring the result of the dead reckoning up to that obtained by observation; always remembering that if a ship is steering east, for example, the allowance FOR THE DIRECTION of the current cannot be the same as would have to be made by a vessel in the same latitudes steering west. *If the allowance for currents he made in the same direction when the ship is steering west as when she is steering east,* IT IS VERY EVIDENT THAT THIS IS DONE TO BRING THE THEORETICAL RESULT INTO LINE WITH THE ACTUAL FACTS. Navigators are often at a loss to account for the great differences between dead reckoning (even when the allowance for currents has been made) and the ship's position as obtained by observation. Believing that they are sailing on a globular surface, nothing presents itself to the mind, but the *usual theories* by which they unsuccessfully endeavour to account for the discrepancy. Did they know that the surface of the ocean is a plane surface (*they* OUGHT TO KNOW THIS), something new would present itself for consideration, theories would be abandoned, and investigation instituted. The result could not fail to be advantageous to navigation generally. In "South Sea Voyages," by Sir James C. Ross, Vol. i, page 96 states:

"We found ourselves every day from 12 to 18 miles observation *in advance* of our reckoning."

Page 27:

"By our observations at noon we found ourselves 58 miles to the *eastward* of our reckoning in two days,"

"Voyage towards the South Pole," by Captain Jas, Weddell, states:

"Feb. 11th, at noon, in lat. 65° 53' South, our chronometers gave 44 miles more *Westing* than the log in three days."

Lieutenant Wilkes says that in less than 18 hours he was 20 miles to the *east* of his reckoning, in latitude 54° 20' South. In "Anson's Voyage round the World," by R. Walter, page 76, the following statement is made:

"It was, indeed, most wonderful that the currents should have driven us to the eastward with such strength; for the whole squadron esteemed themselves upwards of 10 degrees more westerly than this land (Straits of Magellan); so that in running down, by our account, about 19 degrees of longitude, we had not really advanced half that distance."

Captain Woodside, of the American barquentine *Echo,* at Capetown, on 26th June, 1898, reports that on 12th January, 1896, being without observation for two days and going 250 miles a day on a straight course, he expected to be 100 miles south and a long way to the eastward of Gough Island in latitude 40° south, but was startled to find his ship making straight for the island, and barely escaped shipwreck. The *Philena Winslow* was wrecked there 25 years ago, and there are remains of numerous other wrecks.

The fact that in sailing either east or west the currents are allowed *the same way,* proves that the rotundity idea is the factor which effectually debars our navigators from ascertaining a correct solution of the difficulty. Let it be acknowledged that, as the surface of all standing water is level, the world is a plane and not a globe, and investigation may be instituted into the causes of the discrepancies to which we have alluded. But so long as the globular idea prevails, so long will it be impossible for the navigator to arrive at the truth of the matter. I have further weight of evidence on this important branch of our subject, by comparing the theoretical measurements of the supposed "globe" with the distances actually made in sailing. These data, which I now submit, prove clearly to any unprejudiced mind, that the world cannot be the globe of astronomical imagination; but that it is an outstretched circular plane, without axial or orbital motion.

Sir Robert Ball, in his "Story of the Heavens," page 163, informs the reader that:

"The dimensions of the earth are known with a high degree of accuracy."

This writer is recognised as an able exponent of globular hypotheses, and it is generally conceded that what he says may be regarded as correct. Let us now enquire with what high degree of accuracy the dimensions of the earth are known. If the earth be the globe it is generally said to be, it is evident that the further we go south from the equator, the smaller will the circles be, and no circle south of the equator could be equal to that at the equator.

The S.S. *Nithsdale,* of Glasgow, Captain Hadden, sailed from Hamelin Bay, in Western Australia, on 8th January, 1898, arriving at Port Natal on 1st February, 1898, having steamed 4,519 nautical miles. Her log, of which the chief officer, Mr. Boyle (also a passed Master), kindly gave me a copy, shows that she did not make quite a rhomb line track.

Hamelin Bay is in latitude 34° south and longitude 115° 5' east. Port Natal is situate in latitude of 29° 53' south and 31° 4' east longitude. The difference of latitude being so small, we shall not get far out if we take the middle latitude, viz.: 32° south. The difference of longitude is 84° 1' or 4.28 of the complete circle of 360° round the world. Something must be added to the ship's log so as to bring the distance up to the rhomb line track, say too miles; therefore, to find the distance round the world at 32° south it is only necessary to solve the following problem:

As 84° 1': 4,619 nautical or 5,390 statute miles: X. Answer = 23,000 miles, nearly.

This is several thousand miles in excess of what the distance would or could be on a globe. And further south; on a globe, the distance would be less.

In the "Cruise of H.M.S, *Challenger,*" by W. J. J. Spry, the distance made good from the Cape of Good Hope to Melbourne is stated to be 7,637 miles. The Cape is in latitude 34° 31' south and Melbourne in latitude 37° south, the longitude of the Cape being 18° 30' east and Melbourne 145° east. The middle latitude is 35½°. Difference of longitude 126½°, which makes the distance round the world at that latitude (35½°) to be over 25,000 statute miles and as great as the equator is said to be. Thus we see on reliable evidence that the further we go south the greater is the distance round the world. This latter distance is many thousand miles more than the purely theoretical measurement of the world at that latitude south. From the same work, we find the distance from Sydney to Wellington to be 1,432 miles. The middle latitude is 37½°, and the difference of longitude 23° 36', which gives as the distance round the world at latitude 37½° south, 25,500 statute miles! This distance is again greater than the greatest distance round the "globe" is said to be and many thousands of miles greater than could be the case on a globe. Thus, on purely practical data, apart from any theory, the world is proved to *diverge* as the south is approached and not, to *converge,* as it would do on a globe.

Eleven - Fluids

It is in the nature of fluids to be and remain level, and when that level is disturbed by any influence whatever, motion ensues until the level is resumed. Professor Airy tells us, in his "Six Lectures on Astronomy," that "quick-silver is perfectly fluid, its surface is perfectly horizontal." We may add that all fluids are the same, for the reason given by the next writer.

Mr. W. T. Lynn, of the Royal Observatory, Greenwich, in his "First Principles of Natural Philosophy," says: "the upper surface of a fluid at rest is a horizontal plane. Because if a part of the surface were higher than the rest, those parts of the fluid which were under it would exert a greater pressure upon the surrounding parts than they receive from them, so that motion would take place amongst the particles and continue until there were none at a higher level than the rest, that is, until the upper surface of the whole mass of fluid became a horizontal plane."

The *English Mechanic* of 26th June, 1896, says:

"Since any given body of water...must have a level surface, *i.e.,* no one part higher than another, and seeing that all our oceans (a few inland seas excepted) are connected together, it follows that they are all VIRTUALLY OF THE SAME LEVEL."

In March, 1870, the Bedford Canal was chosen to experiment upon with a view of determining whether water was horizontal or convex.

The following argument is taken from the report as printed in the *Field* for 26th March, 1870, and is considered to be *sufficient* and *unanswerable*:—

"The stations appeared, to all intents and purposes, equidistant in the field of view, and also in a regular series; first, the distant bridge; secondly, the central signal; and, thirdly, the horizontal cross-hair marking the point of observation; showing that the central disc 13 ft. 4 in. high does NOT depart from a straight line taken from end to end of the six miles in any way whatever, either laterally or vertically. For, if so, and (as in the case of the disc 9 ft. 4 in. high) if it were lower or nearer the water, it would appear, as that disc does, nearer to the distant bridge. If it were higher, it would appear in the opposite direction nearer the horizontal cross-hair which marks the point of observation. As the disc 4 ft. lower appears near to the distant bridge, so a disc to be really 5 ft. higher would have to appear still nearer to the horizontal cross-hair of the telescope. And therefore it is shown that a straight line from one point to the other passes through the central point in its course, and that a curved surface of water has NOT been demonstrated."

In "Theoretical Astronomy," page 47, it is stated:

"On the Royal Observatory wall at Greenwich is a brass plate, which states that a certain horizontal mark is 154 feet above mean water at Greenwich and 155.7 above mean water at Liverpool."

The difference of the level between Liverpool and Greenwich is thus shewn to be only 1.7 feet. If the world were a globe, the difference of level would be many thousands of feet. It is a common saying that water will find its level, and it is true. If water be dammed back, it will, as soon as released, take the easiest course to where it can find its level. The following from the *Natal Mercury* of 24th October, 1898, fully illustrates this point:

A MOUNTAIN OF WATER

London, Oct. 19 (*Diggers' News* Special). — The steamer *Blanche Rock,* whilst entering the Morpeth Dock, Birkenhead, burst the dock gates. The water inside, which was 8 ft. higher than the level of the river, rushed out with tremendous force. The swirling mass of water damaged the shipping, and beached and sank a number of barges. Two lives were lost.

As soon as the water got to the level of the river, its power would cease.

C. Darwin, in his "Voyage of a Naturalist," page 328, tells us:

"I was reminded of the Pampas of Buenos Ayres, by seeing the disc of the rising sun, *intersected by an horizon* LEVEL AS THAT OF THE OCEAN."

A globe with level oceans would be a new thing in geography!

Twelve - Figure of the Earth

In the "History of the Conflict between Religion and Science," by J. W. Draper, page 153, we are informed that

"An uncritical observation of the aspect of nature persuades us that the earth is an extended level surface which sustains the dome of the sky, a firmament dividing the waters above from the waters beneath; that the heavenly bodies—the sun, the moon, the stars—pursue their way, moving from east to west, their insignificant size and motion round the motionless earth proclaiming their inferiority. Of the various organic forms surrounding man none rival him in dignity, and hence he seems justified in concluding that everything has been created for his use—the sun for the purpose of giving him light by day, the moon and stars by night."

A critical observation of Nature, I may say, persuades an intelligent and unbiased mind that "seeing is believing," and that, therefore, the world is not the globe of modern ideas. Dr. Draper further tells us, on page 156 of his book:

"Many ages previously a speculation had been brought from India to Europe by Pythagoras. It presented the sun as the centre of the system. Around him the planets revolved in circular orbits, their order of position being Mercury, Venus, Earth, Mars, Jupiter, Saturn, each of them being supposed to rotate on its axis, as it revolved round the sun.

"Aristarchus adopted the Pythagorean system as representing the actual facts. This was the result of a recognition of the sun's amazing distance, and *therefore* of his enormous size. The heliocentric system, thus regarding the sun as the central orb, degraded the earth to a very subordinate rank, making her only one of a company of six revolving bodies."

This *speculation* (apt word this) has been shown in the foregoing pages to be without the slightest foundation in fact, and the world shown to be a plane and not a globe.

In "Modern Science and Modern Thought," by S. Laing, the following imaginative proof of the globular figure of the earth is brought forward:

"If, for instance, by travelling 65 miles from North to South, we lower the apparent height of the Pole Star one degree, IT IS MATHEMATICALLY CERTAIN that we have travelled this 65 miles, not along a flat surface, but along a circle which is three hundred and sixty times 65, or, in round numbers, 24,000 miles In circumference, and 8,000 miles in diameter...and that the form of the earth is a perfect sphere of these dimensions."

And on pages 162 and 163 the following is the continuation of the same ridiculous argument:

"Until the Cape was doubled, the course of De Gama's ships was in a general manner southward. Very soon it was noticed that the elevation of the Pole Star above the horizon was diminishing, and soon after the equator was reached the star had ceased to be visible. Meantime other stars, some of them forming magnificent constellations, had come into view—the stars of the Southern hemisphere. ALL THIS WAS IN CONFORMITY TO THEORETICAL EXPECTATION FOUNDED ON THE ADMISSION OF THE GLOBULAR FORM OF THE EARTH."

If we select a flat street a mile long, containing a row of lamps, it will be noticed that from where we stand the lamps I gradually decline to the ground, the last one being apparently quite on the ground, Take the lamp at the end of the street and walk away from it a hundred yards, and it will appear to be much nearer the ground than when we were close to it; keep on walking away from it and it will appear to be gradually depressed until it is last seen on the ground and then disappears. Now, according to the astronomers, the whole mile was only depressed about eight inches from one end to the other, so that this 8 in. could not account for the enormous depression of the light as we recede from it. This proves that the depression of the Pole Star can and does take place in relation to a flat surface, simply because we increase our distance from it, the same as from the street lamp. In other words, the further away we get from any object above us, as a star for example, the more it is depressed, and if we go far enough it will sink (or appear- to sink) to the horizon and then disappear. The writer has tried the street lamp many times with the same result.

Thirteen - Growth of the Earth

R. A. Proctor, in his work "Our place among Infinities," on pages 9 and 10, tells us that the earth was once a mass i glowing vapour,

"capturing then as now, but far more actively then than now, masses of matter which approached near enough, and GROWING by these continual draughts from without...all that is within and upon the earth...are formed of materials which have been drawn in from these depths of space surrounding us on all sides...particles drawn in towards the earth by processes *continuing millions and millions of ages."*

This is written with as much authority as the writer could have had, had he been present when the supposed "spark" was "shot off from the sun." He writes as though he had carefully watched the spark grow bigger, age by age, until it assumed the proportions it had when it "began to cool down." He tells his story as though he had been an eye-witness of all the supposed processes during all the supposed "countless ages" until protoplasm made its appearance and life began to evolve upon the supposed globe. The reader is made to understand, from the "scientific" manner in which the mythical story-teller unfolds his mythical tale, that he, the retailer of the story, carefully watched the evolution of the earth until the time came when the astronomers were able to tell us "without the fear of contradiction" that the earth actually had taken all these millions of ages to evolve into its present form and size. Marvellous, is it not, and how very scientific, to be sure! The reader may pass over the whole of the foregoing extract from the pen of "the greatest astronomer of the age," for there is not one word of truth in it. It is the product of a fertile imagination, nothing more.

The world is much the same as it was in the days of our grandfathers, only the people now are more infidel they were in those days. And since its creation it has not greatly altered, except as it has been altered by the universal flood in the time of that righteous man Noah. The flood disturbed the "strata" of the earth and broke up its layers, hence we find thc bones of men and animals beneath the "crust," which fact causes infidel scientists, who are seeking a proof of the untruth of the Bible, to believe that the earth is many millions of ages old, and therefore not the earth of the creation as recorded in Genesis. The poet Cowper has well said:

"Hear the just law, the judgment of the skies,
He that hates TRUTH shall be the dupe of lies;
And he that WILL be cheated to the last,
Delusions, strong as hell, shall bind him fast."

Fourteen - Gravitation

The "law of gravitation" is said by the advocates of the Newtonian system of astronomy, to be the greatest discovery of science, and the foundation of the whole of modern astronomy. If, therefore, it can be shown that gravitation is a pure assumption, and an imagination of the mind only, that it has no existence outside of the brains of its expounders and advocates, the whole of the hypotheses of this modern so-called science fall to the ground as flat as the surface of the ocean, and this "most exact of all the sciences," this wonderful "feat of the intellect" becomes at once the most ridiculous superstition and the most gigantic imposture to which ignorance and credulity could ever be exposed.

In the "Story of the Heavens," by Sir R. Ball, it is stated on page 82:

"The law of gravitation, THE GREATEST DISCOVERY *that science has yet witnessed.*"

"The law of gravitation WHICH UNDERLIES THE WHOLE OF ASTRONOMY."

Page 101:

"The law of gravitation announces that everybody in the universe attracts every other body with a force which varies inversely with the square of the distance."

"Popular Science Recreations," by G. Tissandier, pages 486 and 487, contains the following:

"Gravitation is the force which keeps the planets in their orbits."

"Every object in the world tends to attract every other object *in proportion to the quantity of matter of which each consists.*"

Professor W. B. Carpenter, in his work "Nature and Professor Man," page 365, says:

"'The laws of light and gravitation,' wrote Mr, Atkinson to Harriet Martineau, 30 years ago, 'extend over the universe, and explain whole classes of phenomena,' and this explanation, according to the same writer, is all-sufficient, '*Philosophy finding no God in nature,* NOR SEEING THE WANT OF ANY.'"

C. Vernon Boys, F.R.S., A.R.S.M., M.R.I., in his paper, "The Newtonian Constant of Gravitation," says:

"G, represents that mighty principle under the influence which every star, planet and satellite allotted course. Unlike any other known physical influence, it is independent of medium, it knows no refraction, it cannot cast a shadow. It is a mysterious power which NO MAN CAN EXPLAIN, OF ITS PROPAGATION THROUGH SPACE, ALL MEN ARE IGNORANT...I cannot contemplate this mystery, at which we ignorantly wonder, without thinking of the altar on Mars" hill. When will a St. Paul arise able to declare it unto us? Or is gravitation, like life, a mystery that 'er be solved?" — *Proceedings of the Royal Institution of Great Britain, March* 1895, *p.* 355.

Professor W. B. Carpenter, in his paper "Nature and Law," published in the "Modern Review" for October, 1890, says:

"The first of the great achievements of Newton in relation to our present subject, was a piece of purely Geometrical reasoning. ASSUMING two forces to act on a body, of which one should be capable of imparting to it uniform motion in a straight line, whilst the other should attract it towards a fixed point in accordance with Galileo's law of gravity, he demonstrated that the path of the body would be deflected into a curve...The idea of continuous onward motion in a straight line, as the result of an original impulsive force not antagonised of by any other — formularised by Newton as his first motion — is not borne out by any acquired experience, and does not seem likely to be ever thus verified. For in no experiment we have it in our power to make, can we entirely eliminate the I antagonising effects of friction and atmospheric resistance; and thus all movement that is subject to this retardation, and is not sustained by any fresh action of the impelling force, must come to an end. Hence the conviction commonly entertained that Newton's first 'law' of motion must be true, *cannot* be philosophically admitted to be anything more than a *probability*....WE HAVE NO *PROOF,* AND IN THE NATURE OF THINGS CAN NEVER GET ONE, OF THE ASSUMPTION OF THE ATTRACTIVE FORCE EXERTED BY THE EARTH, OR BY ANY OF THE BODIES OF THE SOLAR! SYSTEM, UPON OTHER BODIES AT A DISTANCE. Newton himself strongly felt that the impossibility of *rationally accounting* for action at a distance through an intervening vacuum was the weak point of HIS system. All that we can be said to know is that which we learn from our own experience. Now, in regard to the Sun's attraction for the Earth and Planets, WE HAVE NO CERTAIN EXPERIENCE AT ALL. Unless we could be transported to his surface, we have no means of experimentally comparing Solar gravity with Terrestrial gravity; and if we *could* ascertain this, we should be no nearer the determination of his attraction for bodies at a distance. THE DOCTRINE OF UNIVERSAL GRAVITATION THEN, IS A PURE ASSUMPTION."

In "Letters to the British Association," Professor Bernstein says:

"The theory that motions are produced through material attraction is absurd...Attributing such a power to mere matter, which is PASSIVE BY NATURE, is a supreme illusion...it is a lovely and easy theory to satisfy any man's mind, but when the *practical test* comes, it falls all to pieces and becomes one of the most ridiculous theories to common sense and judgment."

The following extracts are taken from "A Million of Facts," by Sir Richard Phillips:

"If the sun has any power, it must be derived from motion; and if acting on bodies at a distance, like Jupiter on his moons, or the Earth on its moon, THERE *MUST* BE AN INTERVENING *MEDIUM* TO CONDUCT ITS MOMENTUM THROUGH ITS SYSTEM."

"It is a principle never to be lost sight of that circular motion is a necessary result of equal action and reaction in contrary directions; for the harmony would be disturbed by variation of distance, if the motion were rectilinear. The same action and reaction are therefore only to be preserved by reciprocal circular motion. NO ATTRACTION AND NO PROJECTILE FORCE ARE THEREFORE NECESSARY. THEIR *invention must be regarded* AS BLUNDERS OF A SUPERSTITIOUS AGE."

"If the bodies came near while moving THE SAME WAY, there would be no mutual REACTION, and they would go together for want of reaction, and NOT OWING TO THAT *MECHANICAL IMPOSSIBILITY* CALLED *ATTRACTION*."

"To accommodate THE *HYPOTHETICAL* LAW OF UNIVERSAL GRAVITATION to the phenomena of the Planets, astronomers have preferred to change the mean density of matter itself; and the Earth, for comparison, being taken at a density of 1,000, to accommodate Mercury to THE ASSUMED LAW, it is taken as 2,585; Venus, 1,024; Mars, 656; Jupiter, 201; Saturn, 103; and Herschel, 218. Consequently, we have the *paradox, that* Jupiter, 1,290 times larger than the Earth, contains but 323 times more atoms. Saturn 1,107 times larger, but 114 times more atoms. Even the Sun, according to these theorists, is but one-fourth the density of the Earth! There may be differences, but chemistry and ail the laws that unite and compound atoms, are utterly at *variance* with so rash and wild an *hypothesis*."

"It is waste of time to break a butterfly on a wheel, but as astronomy and all science is beset with *fancies* about attraction and repulsion, it is necessary to *eradicate them.*

C

A O------|------O B

"It there are two bodies, and it is required to move A to C, the force moving A to C *must* proceed from the side A. Either some impact, or some involvement of a motion towards C, *must* act at A to carry A to C, The modern schools, however, assert that B may move A to C, and A move B to C; and this is *mutual attraction!!* Hence it is necessary to believe that B acts on the side A, where B is *not present;* and that A acts on B on the side B, where A is not present. In other words, A is required to be where it is not, and also be in force at A, so as to B to C! all of which is absurd."

'If in any case A and B approach, it is not because a1 moves B towards itself, or B moves A towards itself, but owing to some causes which affect the space in which A and B i situated; and which causes act on A at A, and on B at B, the statement that A moves B, and B moves A, is ignorance, a is what is meant by attraction. It is also worse than ignorance to justify idleness by asserting that the true cause is indifferent; or to justify ignorance, by asserting that it is unknowable!!"

"This reasoning applies to every species of Attraction, whatever may be the pomposity of equivocal terms in which it is described. Universally, bodies cannot push other bodies towards themselves."

C — — — — — O A B O — — — — — D

"If A and B are said to repel one another, and that B makes A move to C, and A makes B move to D, we have to bear in mind, that while A is moving to C it is in force *only in that direction,* and *cannot,* therefore, be *moving* B towards D. In like manner, while B is moving to D, it is in force *only in that direction,* and *cannot,* therefore, be in force in the contrary direction so as to move A to C," Every species and variety of Attraction and Repulsion are therefore absurd.

"MATTER IS IN ALL CASES *THE CONDUCTOR* OF MOTION. If a body moves, it is because it is the patient of some sufficient momentum of body or matter acting ON the side FROM which the body moves, and only in force in *that direction.*" "Some adopters of attraction, &c., talk, by false analogy, of *drawing,* others of *pulling, lifting,* &c. La Place INVENTS gravitating atoms, and gives them a velocity of 6,000 times that of light, which in some way (known only to himself) performs the work of bringing the body in; others IMAGINE little hooks! As to drawing, pulling, &c., *it behoves them to show the tackle* - the levers, the ropes, etc."

"In spite of all the learning, ingenuity, and elaborations of men, confessedly very able, if there is not and cannot be any action of the nature of attraction, and if the phenomena ascribed to it are local effects of palpable local causes, and if all the phenomena and involvement maybe clearly explained on different principles, then it may be to be lamented that so much ability and character should have been wasted, while a reject for truth and sound reasoning demands that the whole should be FORGOTTEN AS A DREAM, OR DEMOLISHED AS A CARD HOUSE.'

Professor Airy, in his "Lectures on Astronomy," 5th Edition, page 194, informs us:

"Newton was the first person who made a calculation of the figure of the earth on the theory of gravitation. He took the following SUPPOSITION as the *only* one to which his theory could be applied. He ASSUMED the earth to be a fluid. This fluid matter he ASSUMED to be equally dense in every part...For trial of his theory he SUPPOSED the ASSUMED fluid earth to be a spheroid. In *this manner* he INFERRED that the form of the earth would be a spheroid, in which the length of the shorter is to the longer, or equatorial diameter, in the proportion of 229 to 230."

The "New Principia," by N. Crossland, contains the following:

"In ascending a hill we experience a hard struggle, and feel fatigued than when waiting on level ground. Why is this? The Newtonian attributes this to the attraction of gravitation of the earth, against the *pull* of which we have to contend; but if he would be consistent with his theory that the attraction of gravitation *diminishes* inversely as the square of the distance from the centre of the earth, we ought, in defiance of experience, to feel it to be less laborious to ascend a hill than to promenade the same distance on level ground, because as we ascend we *recede from* the centre of the earth; therefore the force of gravitation ought to diminish in a corresponding degree. The Newtonian can only get over this difficulty by a species of scientific quibbling. According to the definition of weight I have given, the solution of the problem is perfectly simple. In ascending a hill a man comes in conflict with the law that the natural tendency of any body is to seek the easiest and shortest route to its level of stability. He chooses the very reverse, and must therefore endure the consequences of acting in opposition to this law. At every step he has to lift his *own weight,* and the higher he mounts the more he feels the influence of the law which he defies. His easiest and more direct course to obey the law of weight is to remain where he is; the next is to descend to a lower level.

"The attraction of gravitation is said to be stronger at the surface of the earth than at a distance horn it. Is it so? If I spring upwards perpendicularly I cannot with all my might ascend more than four feet from the ground; but if I jump in a curve with a low trajectory, keeping my highest elevation about three feet, I might clear at a bound a space above the earth at about eighteen feet; so that *practically* I can overcome the so-called force (pull) at the distance of four feet, in the proportion of 18 to 4, being the *very reverse* of what I ought to be able to do, according to the Newtonian hypothesis.

"Again, lake the case of a shot propelled from a cannon. By the force of the explosion and the influence of the reputed action of gravitation, the shot forms a parabolic curve, and finally falls to the earth. Here we may ask, why — if the forces are the same, viz., direct impulse and gravitation — does not the shot form an orbit like that of a planet, and revolve round the earth? The Newtonian may reply, because the impulse which propelled the shot is *temporary; and* the impulse which propelled the planet is *permanent.* Precisely so: but why is the impulse *permanent* in the case of the planet revolving round the sun? What is *the cause* of this permanence?

"We are asked by the Newtonian to believe that the action of gravitation, which we can easily overcome by the slightest exercise of volition in raising a hand or a foot, is so overwhelmingly violent when we lose oar balance and fall a distance of a few feet, that this force, which is imperceptible under usual conditions, may, under extraordinary circumstances, cause the fracture of every limb we possess? Common-sense must reject this interpretation. Gravitation does not furnish a satisfactory explanation of the phenomena here described, whereas the definition of weight already given does, for a body seeking in the readiest manner its level of stability would produce precisely the results experienced. If the influence which kept us securely attached to this earth were identical with that which is powerful enough to disturb a distant planet in its orbit, we should be more immediately conscious of its masterful presence and potency; whereas this influence is so impotent in the very spot where it is supposed to be moat dominant that we find an insurmountable difficulty in accepting the idea of its existence. Fortunately for our faculty of locomotion, the Newtonian hypothesis may rejected as a snare and a delusion.

"It is quite amusing to watch Newtonians and Darwinian floundering about in their attempts to expound the mysteries of creation. Their theories are as ridiculous as the fashion which once prevailed for Della-Cruscan poetry, and they ought to be treated with equal severity.

"It seems quite possible that during the last two hundred years we have been living in a sort of scientific fool's paradise, and that universal gravitation is a gigantic Newtonian mare's

"As a theoretical scientific guide we must give up Sir Isaac Newton as useless and misleading, and allow his reputation to retire into private life.

"In *Knowledge* of the 17th and 24th Feb., 1882, there appeared a discourse on *The Birth of the Moon by Tidal Evolution,* by Dr. Ball, the Astronomer Royal for Ireland, which I should say is *without exception,* the most delusive and absurd contribution ever made to so-called science. At one time I thought that "Parallax," who told us that the earth was a flat plane like a plate, was the most misguided man in the kingdom, but I now believe that he is quite entitled to take rank in scientific wisdom, and to at down on an equality with the Astronomer Royal of Dublin."

I have quoted at length on this important matter, and the evidence here produced, besides very much more in the same direction, for which I have not the space here, shows clearly that THERE IS NO SUCH FORCE AS GRAVITATION IN EXISTENCE ANYWHERE.

One of the world's so-called *great thinkers,* J. S. Mill, is quoted in Professor Carpenter's "Nature and Man," page 385, as saying:

"Although we speak of a man's fall as caused by the slipping of his foot, or the breaking of a rung (as the case may be the *efficient cause* IS THE ATTRACTIVE FORCE OF THE EARTH, which the loss of support to the man's foot brings into operation."

If a man is not "deeper" than to believe what this "deep" thinker has left on record in this matter; if he has no more brain power than to accept the foregoing statement, I would strongly advise him to cease thinking altogether, and thus save the few brains he has. It is simply astounding that men, who in business matters are sharp enough, are as dull as bricks and as credulous as children when the awe-inspiring subject of gravitation, "that grand masterpiece of astronomy," is the theme. To ask the reason why, or to venture to suggest that the assumptions of the "learned" require some sort of proof to back them up, never seems to strike moderns who believe in this monstrous humbug. A. Giberne, in "Sun, Moon, and Stars," page 27, says:

"If the sun is pulling with such power at the earth and all her sister planets, why do they not fall down upon him?"

A very proper question, truly. And when this question is propounded to astronomers, they cannot give an answer worth recording. They simply do not know how to answer the question without stultifying their commonsense. But the above writer thinks it can be answered, so says:

"Did you ever tie a ball to a string and swing it rapidly round and round your head? If you did, YOU MUST HAVE NOTICED THE STEADY OUTWARD PULL OF THE BALL."

The "steady outward pull of the ball" clearly implies that the ball has *intelligence,* and knows just what to do so prevent its hitting the head of the operator. The "outward pull" of a ball which is fastened to the hand of the operator by a string, is clearly impossible. If the operator ceased to impel it round and round his head by the mechanical attachment and the power he exerts in swinging it round, the ball would seek its level of stability and fall to the ground. And, as this illustration is used to teach what gravitation is, and how it acts, we shall just follow the illustration to its logical issue, and see where the theory is. The illustration implies that BETWEEN ALL THE BODIES IN THE UNIVERSE. THERE IS A CONNECTING LINK, which keeps the "body" that attracts attached to the "body" that is attracted. This connecting link, in the case of the ball, is the string, Now, we could readily understand gravitation if this, illustration conveyed to us by the ball and the string were a correct representation of fact. But, we very naturally ask, what is the connecting link? Of what does it consist? And of what do all the connecting links between the sun and the myriad orbs of heaven consists would not the "strings" get somewhat entangled? Has this connecting link ever been observed anywhere? The answer to these pertinent questions is that THERE IS NO CONNECTING LINK in existence. When the "missing link" is produced, we are prepared to admit all the gravitation theorists teach on the subject. Until then we shall continue to regard it as the myth it undoubtedly is. But we are not done with the illustration yet. The "ball and string" device sets forth that the "body" that attracts is not only connected with the "body" attracted, but that the former IS THE MOTIVE POWER OF THE LATTER. — that the sun is the power which compels the earth to | revolve round it, even as the motive power of the ball is the exertion of the hand of the operator. Without the connecting link the earth would fall (according to the astronomers) in a rectilinear path for ever. But what these wise men do not see, and which is a necessary part of the theory, as represented by the ball and string idea, is that the motive power also must come from the sun. Without this motive power and the connecting link, the whole of the theory falls to pieces. THERE IS NO MOTIVE POWER IN THE SUN TO CAUSE THE EARTH TO REVOLVE AROUND IT, AND THERE IS NO CONNECTING LINK BETWEEN THE SUN AND THE EARTH TO KEEP THE LATTER IN ITS POSITION, *consequently the theory of universal gravitation has no existence in fact.* "He who *cannot* reason is a fool; he who *will not* reason is a bigot; he who *dares not* reason is a coward; but he who *can* and *dares* to reason is a MAN."

If the reader can and dares to reason, let him reason this matter out and discover whether astronomy as drummed into children's heads at school, and vauntingly displayed, with many pictures, from public platforms, has one inch of standing ground, or one *reason* to offer as an apology for its further existence and power to befool mankind longer. These are strong statements, but not stronger than the facts warrant.

"The Story of the Heavens," by Sir Robert Ball, is not only an authoritative treatise, which it is, coming from such a recognised exponent of the "science"; but a fulsome account of general principles and details in popular form. As a literary production, it possesses considerable merit, and its good English entitles it to the respect and consideration of all its readers. But as a contribution to science, it is the most absurd and unreasoning conglomeration of nonsensical and impossible ideas I have ever read.

On page 110 of this book, we read that

"Kepler *found* that the movements of the planets could be explained by *supposing* that the path in which each one revolved was an ellipse. This in itself was a DISCOVERY of the most commanding importance."

To explain anything by a supposition, and then to label the supposition a *discovery* is ridiculous in the "domain of science" and a marvel of literary ingenuity.

On the same page, the first law of planetary motion is enunciated in these words, "each planet revolves around the sun in an elliptical path, having the sun as one of the foci," and on page 112 the ellipse is shown with the sun in one focus. Throughout the book, however, the other focus is not mentioned, and it is very evident from the diagram that if the sun were of sufficient power to retain the earth in its orbit when nearest the sun, when the earth arrived at that part of its elliptical path farthest from the sun, the attractive force (unless very greatly increased) would be utterly incapable of preventing the earth rushing away into space "in a right line for ever," as astronomers say.

On the other hand, it is equally clear that if the sun's attraction were just sufficient to keep the earth in its proper path when farthest from the sun, and thus to prevent it rushing off into space; the same power of attraction when the Earth was nearest the sun would be so much greater, (unless the attraction were very greatly diminished) nothing would prevent the earth rushing towards and being absorbed by the sun, there being no counterbalancing focus to prevent such a catastrophe! As astronomy makes no reference to the increase and diminution of the attractive force of the sun, called gravitation, for the above necessary purposes, we are again forced to the conclusion that the great "discovery" of which astronomers are so proud is absolutely non-existent. The law of dynamics, assisted by geometry, makes it, as the learned say, "mathematically certain" that no such force as gravitation exists anywhere-in the universe. As another has well said, its invention must be regarded as a blunder of a superstitious age.

If the earth were the globe of astronomical invention, and if gravitation were needed to keep it in its path around the sun, it is easily seen that gravitation must be circular, as then and then only, would the attraction be equal in every part of the path, and so cause the earth to describe an exact circle throughout the year. Astronomers say that the earth moves and not the sun. And that this movement of the earth causes the seasons. And further, that the movement of the sun which we see is *really* caused by the movement of the earth. If, therefore, the sun *appears* to make an exactly circular path every day of the year, there might be some ground for the astronomers' supposition of gravitation. That the sun's path is an exact circle for only about four periods in a year, and then of only a few hours — at the equinoxes and solstices — completely disproves the "might have been" of circular gravitation, and by consequence, of all gravitation.

It has long been pointed out that gravitation, if it existed at all, must be circular, as the following from Drapers' "Conflict between Religion and Science," page 168, shows:

"Astronomers justly affirm that the book of Copernicus, 'De Revolutionibus,' changed the face of their science. It incontestably established the heliocentric theory. It showed that the distance of the fixed stars is infinitely great, and that the earth is a mere point in the heavens. Anticipating Newton, Copernicus imputed gravity to the sun, the moon and heavenly bodies, hut he was led astray by assuming that the celestial motions must be circular. Observations on the orbit of Mars, and his different diameters at different times, had led Copernicus to this theory."

That the paths of the orbs of heaven are not exactly circular disproves the theory of gravitation entirely.

It is impossible to make a ball tied to the hand with a string revolve in an elliptical path, circular motion being possible. So we may consign the illustration, together with the thing it is intended to illustrate, into oblivion.

The volume already quoted, "Sun, Moon, and Stars," stated, on page 73, that

"Comets obey the attraction of the sun, *yet he appears to have a singular power of driving the comets' tails away from himself.* For, however rapidly the comet may be rushing round the sun, and however long the tail may be, IT IS ALMOST ALWAYS FOUND TO STREAM IN AN OPPOSITE DIRECTION FROM THE SUN."

Here we have an acknowledged failure of the law of gravitation, which is said to be universal. Now comes a declaration which supports my contention that gravitation is non-existent.

In "Science and Culture," by Professor T. H. HUXLEY, page 136, the following statement is made:

"If the law of gravitation EVER FAILED TO BE TRUE, EVEN TO THE SMALLEST EXTENT, for that period, the CALCULATIONS OF THE ASTRONOMER HAVE NO APPLICATION."

After such an "authoritative" declaration, we may well dismiss the subject, and we are fairly entitled to conclude, with such a consensus of evidence against the commonly received "view" of gravitation, together with the application of the principles of sound logic, that GRAVITATION HAS NOT AND NEVKR HAD ANY EXISTENCE, and the idea of such a force must be relegated to the limbo mythology.

Fifteen - Geology

In "Geology," by Skertchley, page 101, it is confessed:

"So imperfect is the record of the earth's history as told in the rocks, that we can ne'er hope to fill up completely all the gaps in the chain of life. The testimony of the rocks has been well compared to a history of which only a few imperfect volumes remain to us, the missing portions of which we can only fill up by conjecture. What botanist but would despair of restoring the vegetation of wood and field from the dry leaves that autumn scatters? Vet from less than this the geologist has to form all his ideas of past floras. Can we wonder then at the imperfection of the geological world?"

The Vice-President of the Royal Geographical Society of Ireland holds that this, the only earth, was made during six successive periods, corresponding to six series of rocks and that particles of mud and sand deposited by rivers in sea bottoms could only become rocks of a heterogeneous mixture, but never such as the primary with sub-divisions, having" each its own marked peculiarities. In his "Errors of Geologists," page 15, he says:

"Neither the brown gneiss, nor the primary red sandstone, nor the yellow quartz rock, nor the gray mica slate, nor the blue limestone. Not one band out of all these could be formed out of the river sediment coming down from the pre-existing continents, because not one of them has mixed particles. The quarts rock has no lime, the limestone is purely crystalline, &c."

Although the deepest mine in the world is only a few thousand feet down, the assertions of geologists that they know what underlies the "crust" of the earth to a depth of 4,000 miles, are received as though they had actually been down making a personal inspection and favoured the world with the result of their researches. Sir D. Brewster, in his "More Worlds than One," says:

"The proportional thickness of these different formations have been *estimated* by Professor Phillips as follow, but the numbers can be regarded only as *a very rude estimate:* —Tertiary 2000 feet, Cretaceous 1000 feet, Oolite and Lias 2500 feet, New Red Sandstone 2000 feet, Carboniferous 10,000 feet, Old Red Sandstone 9000 feet, Primary Rocks 20,000, equals NINE MILES nearly."

"On these ASSUMED data they founded different theories of volcanoes."

"It is TAKEN FOR GRANTED that many of the stratified rocks were deposited at the bottom of the sea by the same slow processes which are now going on in the present day."

Almost needless to remark that whatever speculations nave nothing better than "taken for granted" to support I them, must be rejected as purely fanciful and utterly incapable of proof. Geologists are very fond of parading their knowledge (?) of what they are pleased to term the "glacial period" of the earth's history. Sir R. Ball writes a book on "The Cause of an Ice Age." But he vitiates the entire volume by stating;

"I have found it necessary to ASSUME the existence of several ice ages."

He then goes on to endeavour to prove his assumption to be correct by stating;

"In fact it might almost be said that the astronomical theory (of accounting for ice ages) must be necessarily true, as it is a strictly mathematical consequence FROM THE LAWS OF GRAVITATION."

We have already seen that this magical, indefinable, what-do-you-call-it influence has no existence. We may, therefore, reject the learned writer's "mathematical consequence" as a myth.

In his "Second Appeal to Common-sense from the Extravagance of some Recent Geology," Sir H. H. Howorth, K.C.I.E., M.P., F.R.S., F.G.S. says:

"One of the chief objects of this book is to show that the Glacial theory, as usually taught, is not sound; but that it ignores, and is at issue with, the laws which govern the movements of ice, while the geological phenomena to be explained refuse to be equated with it. This is partially acknowledged by the principal apostles of the ice theory. They admit that ice as we know it in the laboratory, or ice as we know it in glaciers, acts quite differently to the ice they postulate, and produces different effects; but we are bidden to put aside our puny experiments which can be tested, and turn from the glaciers which can be explored and examined, to the vast potentiality of ice in shape of portentous tee-sheets beyond the reach of empirical tests, and which we are told acted quite differently to ordinary ice. That is to say, they appeal from sublunary from experiments to *à priori* argument drawn from a transcendental world. Assuredly this is a curious position for the champions of uniformity to occupy."

"I hold that the Glacial Theory, as ordinarily taught, is based, not upon induction, but upon hypotheses, same of which are incapable of verification, while others can be shown to be false, and it has all the Infirmity of the science of the Middle Ages. This is why I have called it a Glacial Nightmare. Holding it to be false, I hold further that no theory of modern times has had a more disastrously mischievous effect upon the progress of Natural Science."

"I not only disbelieve in, but I utterly deny, the possibility of ice having moved over hundred of miles of level country, such as we see in Poland and Russia, and the prairies of North America, and distributed the drift as we find it there. I further deny its capacity to mount long slopes, or to traverse uneven ground. I similarly deny to it the excavating and denuding power which has been attributed to it by those who claim it as the excavator of lakes and valleys, and I altogether question the legitimacy of arguments based upon a supposed physical capacity which cannot be tested by experiment, and which is entirely based upon hypothesis. This means that I utterly question the prime postulate of the glacial theory itself."

In the *Scientific American Supplement* of 10th September, 1898, in an article on "Glacial Geology in America," by H. L. Fairchild, the following is stated:

"The cause of the glacial period remains quite as much a mystery as it was in 1840. A large body of fact has been collected, but it points in different directions. *Every person has entire liberty of opinion.* MOST GLACIALISTS HAVE NO OPINION AT ALL UPON THIS SUBJECT."

The reader need not trouble to have any opinion on the subject, for *there never was a glacial period in the history of the world.* We challenge the whole scientific world to prove the romance.

A. McInnes, in his paper "The Flood and Geology," says:

"Next, how was the flood caused? Moses says by the opening of the netting (not windows) of heaven to pour down ra.in, and by the opening of the fountains of the great abyss of waters. What deplorable ignorance prevails regarding the true Constitution of the universe. The old pagan delusion of Pythagoras is now generally believed in opposition to common sense, reason, and God's own revelation—that men are now living on an impossible large ball of land and water, flashed above and round the sun more quickly than a thunderbolt. Thus apostle's prediction is fulfilled, that men in the last days would not endure sound doctrine, but would give heed to Cables. As of old so now, 'they glorify not God, but have become vain in their reasonings and their heart is darkened. Professing themselves wise they have become fools.'—Romans i, 21.

"We have God's own revelation — Job xxxviii, — manifestly opposed to the fables now falsely called science. God asks Of Job — 'Where wast thou when I laid the foundation (Heb. fixed) of the earth?' Where has the earth or land been fixed? 'He has founded it upon the seas, and established it upon the floods.' — Ps. xxiv. 2. 'The earth *standing* out of the water and in the water.' — 2nd Peter iii. 5. Thus the land does not, as is assumed without proof by modern astronomers, contain the sea; but the sea contains the land, and is the great abyss out of which the dry land appeared at God's creative word, — Gen. i. 9. Likewise, the Antarctic icebergs surround the sea on every side, utterly baffling all attempts of navigators to proceed further south. 'Who shut up the sea with doors, and prescribed for it my decree, and set bars and doors, and said: "Hitherto shalt thou come, hut no further; and here shall thy proud waves be stayed,"?' — Job xxxviii. 8. Next, as to the structure of the earth it was asked: (v. 5) 'Who has set its layers?' or, 'laid its measures?' 'Or, who stretched out a measuring line upon it?' 'On what are its bases (or sockets) sunk?' 'Or who laid down its keystone rock?' This rendering is precisely according to the Hebrew. Now, does not the fifth verse plainly declare that the earth's strata or layers were arranged by God himself, and not according to suppositions of modern geologists? The layers are found to be set with the regularity and exactness of the stones of a house, and as if the builder's measuring line had been used. The unstratified or key-stone rock, whether basalt or granite, lowest en the sea, but above are the various beds according to density, such as sandstone, slate, limestone, coal, chalk, clay, with sand, gravel or soil on the surface. How the all-wise God did 10,000 years ago produce by His almighty word the vast construction of the earth's interior in such wonderful as the geologists foolishly suppose ways are not as man's nor His thoughts as ours. He also beginning made all the various kinds of animals, not according to a slow process of growth or development; but the birds and fishes on the fifth day, beasts, creeping things and man on the sixth day, each kind separate from the other, contrary to that be atheistic supposition of evolution; and the day limited by the evening and morning, 12 hours. 'Are there not 12 hours in the day?' asked the Lord.

"Thus the whole mighty mass of rock, stratified and unstratified, has been made to float upon the unfathomable waters, yet as securely fixed as a ship in a Liverpool dock. The bases of the earth are so sunk as to make it immoveable forever. Man is challenged to tell how. 'Upon what are its bases sunk?' Job 38. "He founded the earth on its basis; it is not moved forever and ever." — Ps. 104, 5, Now, why can an iron ship float, though that metal is seven times heavier than water? Because, chiefly of the shape. Bui the heaviest rock is only three times the weight of wafer. Then consider the tremendous buoyancy of the ocean causing some substances to float on the surface, and others to sink only to a certain depth. The earth, its density decreasing from the foundation rock upwards to the soil of the surface, is sunk to a depth several miles in the sea, yet so as to have a dry surface, and shores on a level with the surrounding waters. It consists of four continents of an irregular and somewhat triangular shape, stretching out from the central north, thousands of miles towards the icy barriers of the far south, against which winds and waves rage in vain. The continents are connected by sub-marine rocky beds, varying in depth, whilst the Arctic and Antarctic oceans are found to be unfathomable.

"The flood, as we have seen, was caused by the opening of the netting of heaven and the fountains of the abyss. The heaven or sky is an expanse for the clouds, strong as molten mirror.' - Job 37, 18; and was made on the second day of creation to separate the waters above from the waters below. 'Hast thou come to the springs of the sea?' asks God - 38, 16. It was formerly the opinion of Christian writers that these springs or fountains are in the central north, confined by the impenetrable walls of ice, which were broken down at the flood. However, when

Noah had entered the ark, from heaven and the abyss rushed the waters to fulfil God's purpose to destroy the earth with its inhabitants. Hence, the rending of rocks, the shattering of hills, the breaking up of the earth's strata, the piling of mass upon mass, wherein were buried annuals and plants to be dug up many centuries afterwards. All lands were filled with the wreck of the old world — a terrible warning to all future ages against the commission of unrighteousness.

"And, let it be noted that the petrifaction of fossils is not surprising, seeing that the earth was wholly sunk under the waters for a whole year. Even geologists confess that the degree of petrifaction is no proof of the antiquity of a fossil. 'The mere amount of change, then, which the fossil has undergone, is not by any means a proof of the length of time that has elapsed since it was buried in the earth; as that amount depends so largely on the nature of the material in which it was entombed, and on the circumstances that have since surrounded it.' — Jukes, p. 190.

"Then, what was the origin of the rocks, indeed of the entire earth? Aqueous, according to Genesis. 'In the beginning of God's framing the heavens and the earth, the earth was in loose atoms and empty.' (Hebrew) Where were the loose atoms? In the abyss of waters; and God on the. third day of creation consolidated all into rocks, stratified and unstratified, causing the land to appear.

"But why is man not found as a fossil embedded among the rocks, as are the animals? The answer is not difficult. Before the flood man was not so prolific as now. During the 1656 years of the old world there were, according to Moses, only ten generations counting from Adam to Noah; and Noah during' 600 years had only three sons. However, let us reckon approximately the antediluvian population, allowing eight children to each couple. 1st generation, 2; 2nd generation, 8; 3rd generation, 32; 4th generation, 128; 5th generation, 512; 6th generation, 2,048 7th generation, 8,192; 8th generation, 32,768; 9th generation. 131,072; 10th generation. 524,288, The sum is 699,050; and the whole human population before the flood might not amount to one sixth of the population of London. But it remembered that mankind in the old world dwelt in Asiatic Turkey, speaking the same language, and it was not till after Noah's death that the dispersion from Babel over the earth took place. Asiatic Turkey contains at present fifteen million human beings, and there only could fossilised man be found. To what extent, if at all, has that country been geologically examined?

"Is it possible to deliver men from the spell and sorcery of 'great names?' If only a fable or lie is called scientific, and, fathered by a writer reputed a 'great man,' how many thousands believe at once without proof? Is it not as hard to turn men from the worship of their fellow-worms, as to turn a Hindoo from the worship of sticks and stones? The scientific favourites of newspaper scribblers are larded over reputation of greatness is attained; and to argue against scientific fictions is only to provoke silly jesting or astonishment at the presumption of daring to differ from the scientific slave-drivers. Will any of their slaves of science dare be free, or use their common-sense?"

"Is geology not a tissue of suppositions from beginning to end? Let us see. How do the Geologists manage to get dupes? Some disguised infidel who has had sufficient influence to obtain a professorship in a college writes a book about the Creation, in which he attempts to prove to the entire satisfaction of atheistic journalists that the world made itself without the help of God at all. Of course the blasphemous character of the book is carefully veiled, lest soft-headed religionists take alarm, and the book does not sell. Perhaps even a pious whine is dropped so that the work of Judas may be done more effectually; and the author is reputed so very great a man, for all the newspapers say it. By way of preface astronomy is appealed to as a science so well-established that none but fools object to it; therefore, the reader must imagine all the vast continents and oceans making up a ball no larger than the school room globe. Next he is assured that recent researches in science have proved that those lights, the sun, moon, and stars, consist of the very same constituents the earth and sea, as well as the nebulae, which science supposes to be clouds of glowing gas. So all these must have had a common origin, and, therefore, the simpleton must next imagine the school room globe along with sun, moon and stars, changed into a quantity of fiery gas. In the beginning — how many million years ago science cannot yet decide — was gas, is the dogma of Geology. But he dare not ask about the origin of the gas itself. Then the mesmerist requires him to suppose that all the fiery mass very conveniently began to cool, particularly a quantity in the centre, which also whirled about until it became the sun."

"The victim of duplicity is next to suppose that other quantities also cooled until they changed into planets. Especially one quantity went on cooling until it very conveniently became the earthball with a rocky crust, and though on fire originally, yet a portion of it changed into all the oceans and seas. 'In the study of science,' says Dr. Dick in his book on Geology, 'one is permitted to suppose anything if he will but remember and acknowledge to others that he only makes suppositions; will give reasons to show that his suppositions may be true, and be ready at any time to give up his suppositions when facts go against them. The last of these two suppositions, namely, the gradual cooling of the world from a state of intense heat, is often made by those who wish to form to themselves a notion of how the rocks and rivers, mountains and plains of the world have been brought to exist as they are.' p. 10. Can the foolish Geologists, instead of making these absurd suppositions, not believe the fact that God made the world as stated on God's own authority? Instead, however, of opening their eyes they further suppose that despite the cooling, as much fire remained inside the ball as heaved up the rocky crust into mountain chains, whilst the waters went on channelling and levelling so as to make all the river and ocean beds. Then the rivers would carry down to lakes and seas matter containing animal and vegetable remains to form sediment, which we must suppose hardened after millions of years into rocks, especially the stratified ones, the unstratified rock being supposed due to the original fire. All these atheistic suppositions are expressed in words of Greek origin so as to amaze the gaping simpleton. The rocks immediately above the unstratified are called metamorphic. Next in ascending order are the palaeozoic or primary, the mesozoic or secondary, the cainozoic including the tertiary and quaternary. The guesses about fossils make up Palaeontology.

"Now, let it be observed that not one of these suppositions is even probable. Who ever saw gas changed into granite, or a fiery vapour into water, or a river channel its own bed? Is there within the memory of mankind one considerable mountain more or less on the earth — notwithstanding volcanic eruptions and earthquakes — one considerable county more or less, or what continent has materially changed its shape? What do fossils prove? The following is a confession from Skertchly's Geology, p. 101:— 'So imperfect is the record of the earth's history, as told in these rocks, that we can never hope to fill up completely all the gaps in the chain of life. The testimony of the rocks has been well compared to a history of which only a few imperfect volumes remain to us, the missing portions of which we can only fill up by conjecture. What botanist but would despair of restoring the vegetation of wood and field from the dry leaves that Autumn scatters? Yet from less than this the Geologist has to form all his ideas of past floras. Can we wonder then at the imperfection of the geological world?' Indeed it is confessed that the age of a fossil is not determined by the degree of its petrification, but by the age of the rock in which it is imbedded; and the age of the rock by its position among the strata. Have men in these last days become so silly that with old bones and stones, and foot-marks, they may be led to deny the very God that made them? But was not this folly foretold ages ago by the inspired Hebrew prophets?

"Each layer of rocks is supposed by Geologists to have occupied an indefinite number of millions of years, and the age of the earth is still more a mystery to them. Professor Thomson, who is a scientific dictator, has, however, announced that the solidification of the earth could not have taken less than 20,000,000 years, and not more than 400,000,000 years, and so that the date of the world's beginning is somewhere between these two numbers. Some time ago Geologists proved from scientific data (to their own entire satisfaction and that of their dupes), that the earth is a ball of liquid fire with a thin crust of rock, so that at a depth of 25 miles the rocks must melt, and at 150 they would go off in vapour. (Dr. Dick's Natural History, p. 12). But Professor Thomson has found out that those suppositions do not square with the supposition of gravitation, and accordingly he supposes rather that the mass of the earth can" not be much less rigid than a globe of steel of the same size would be, yet that there must be some quantity of the fiery liquid left in the interior, enough at least to cause earthquakes and volcanic eruptions. What tinkering the imaginary globe of the astronomer needs?

"Some geologists, such as Jukes, are not certain whether the earth was a molten mass at first, and whether granite is of igneous or aqueous origin. Formerly rocks were classified as primary, transition, secondary, tertiary, recent, but now by a new arrangement the transitionary rocks are denied any place in the series. Jukes says that he holds views with regard to the Devonian period which differ from those taken by most geologists, and that the question is hardly yet settled,' p. 203. Also, regarding the stratified rocks, he observes, 'that at one time it was thought that there was some essential distinction in the nature of these rocks, and their mode of formation. It is now known that the primary rocks when first formed were exactly like the corresponding secondary and tertiary,' p. 202. Indeed, is there anything certain about geology except that it is disguised atheism denying God the Creator?

"Geologists profess to prove extinct species. Of course they can produce large bones to show that at one time there were large elephants and lizards, but are big dogs not dogs as really as little ones? Is it a fact according to Moses, there were human giants before the flood, and that, since the lower animals have degenerated in size and age as well as men, need not surprise this nineteenth century of crime and infidelity. But the trick of comparative anatomy is to claim with an old bone the power of reproducing the sketch of the entire animal, though formerly unknown. If the monkey had been unknown to Darwin and the scientists, would they have been able by seeing one hand only, to tell that that beast has four hands? If zoologists think the serpents once had wings or feet, let them read Genesis iii. 14 — 'On thy belly shalt thou go.' Let scientists ere concluding that any kind of animal has become extinct consider the words of Jukes himself: 'As *all the truth* about anything whatever is absolutely unattainable by us, it would only lead us astray if we required it from Geology, or reasoned as if we had attained it,' p. 202. But recently the existence of the gorilla became known. What of the leviathan, the swift serpent, the crooked serpent, the dragon that is in the sea,—Isa. xxvii. Is it not chiefly the fossilised bones of the sea serpent that geologists are exhibiting as the remains of extinct species of a vast size? No wonder the present existence of the leviathan is so eagerly denied."

S. Laing, in his "Modern Science and Modern Thought," page 27, informs us that

"The total thickness of *known strata* is about 130,000 feet, or 25 miles...of this, about 30,000 feet belong to the Laurentian, which is the oldest known stratified deposit, 18,000 to the Cambrian, and 22,000 to the Silurian. These form together what is known as the Primary or Palaeozoic Epoch."

Mr. Laing is very careful to omit the names of those who *know* strata for a depth of 25 miles. Can it be that he has been down there himself? If so, we may expect to have further revelations as to the contents of the bowels of the earth. But no, he cannot have been there, for he tells us a little further on (page 37):

"At this rate of increase water would boil at a depth of 10,000 feet, and iron and all other metals be melted before we reached 100,000 feet."

We are thus satisfied that the gifted author was not actually there, *or he would have been melted in company with "iron and all other metals."* This is a relief, and enables us to at once and for ever dispose of his wild theories as baseless assumptions. In a certain case before the Magistrate, the culprit hardly liked to say that the witness against him was telling a He, so he mildly said that the witness was "handling the truth very carelessly." When Mr. Laing has the impertinence to tell us what lies below the surface of the earth for a depth of 25 miles we are bound to say that he handles the truth in a careless and most reprehensible manner.

With the usual unqualified manner for which scientists have become famous, Mr. Laing goes on to say:

"Reasoning from these *facts,* ASSUMING the rate of change in the forms of life to have been the same formerly...Lyell has arrived at the conclusion that Geology requires a period of not less than 200,000,000 of years to account for the phenomena which it discloses."

To reason from *facts* and then to assume something which in its very essence is utterly incapable of proof, is bad enough; but to mis-call fictions facts and then to add on to them whatever assumption is necessary to maintain the *result* in keeping with the theory with which the start was made, is so atrocious that we are again forced to the conclusion that Geologists are lost in the fogs of their own creation, and cannot find their way through the millions of ages of their own imagination, to anything having the remotest bit of truth in it. Once more, and I have done with Mr. Laing and his Geology. He informs us in the work already referred to that:

"The law of gravity, which IS THE FOUNDATION OF MOST OF WHAT WE CALL THE NATURAL LAWS OF GEOLOGICAL ACTION has certainly prevailed, as will be shown later, through the enormous periods of geological time and far beyond this WE CAN DISCERN IT OPERATING in those astronomical changes by which cosmic matter has been condensed into nebulas, nebulae into suns throwing off planets, and planets throwing off satellites, as they cooled and contracted."

The laws of geological action being based on a myth—the law of gravitation, Geology itself may be "thrown off into space" without any ill effects being felt anywhere.

GEOLOGY and ASTRONOMY as at present taught by the schoolmen are nothing more than fables.

Hear what *The Future* of February, 1892, says:

"Astronomers are very fond of boasting of the wonderful exactness of their science, and that it is based on the principles of incontrovertible mathematics; and of ridiculing astrology as a *pseuda*-science. The exactness belongs to practical and not to theoretical astronomy. For example, when the writer learnt the principles of astronomy at school, he was taught that the Sun was exactly 95 millions of miles from the earth; now-a-days astronomers say that this was an error, and that the Sun is only 92 millions of miles distant. Newton made the Sun's distance to be 28 millions of miles, Kepler made it 12 millions, Martin 81, and Mayer 104 millions! Dr. Woodhouse, who was professor of astronomy at Cambridge about fifty years ago, was so candid as to admit the weakness of the Newtonian speculations. Woodhouse wrote: 'However perfect our theory, and however simply and satisfactorily the Newtonian hypothesis may seem to us to account for all the celestial phenomena, yet we are here compelled to admit the astounding truth that if our premises disputed and our facts challenged, the whole range of astronomy does not contain the proofs of its own accuracy.'"

Sixteen - The Horizon

According to tables of curvature compiled to suit the mathematical factors and tentative formulas employed in the imaginary geodetic operations, which have from time to time been conducted in observatories, the horizon of an observer is distant or near according to the greatness or otherwise of his elevation above the surface of the supposed globe. If he stands 24 feet above sea level, he is said to be in the centre of a circle which bounds his vision, the radius of which in any direction, on a clear day, is six miles.

A local gentleman tells me that he has watched a boat-race in New Zealand, seeing the boats all the way out and home, the distance being 9 miles from where he was standing on the beach. I have seen the hull of a steamer with the naked eye at an elevation of not more than 24 feet, at a distance of 12 miles, and in taking observations along the South African coast, have sometimes had an horizon of at least 20 miles at an elevation of 20 feet only. The distance of the horizon, or vanishing point, where the sky appears to touch the earth and sea, is determined, largely by the weather, and when that is clear, by the power of our vision. This is proved by the fact that the telescope will increase^ the distance of the horizon very greatly, and bring objects into view which are entirely beyond the range of vision of the unaided eye. But, as no telescope can pierce a segment of water, the legitimate conclusion we are forced to arrive at, is that the surface of water is level, and that, therefore, the shape of the world cannot be globular, and on such a flat or level surface, the greater the elevation of the observer, the longer will his range of vision be, and thus the farther he can see.

Seventeen - On the Term "Level"

Advocates of the globular form of the world often fall back on the meaning of the term "level," affirming that a level surface means an even surface and not a horizontal or flat one. That is to say that a convex surface if free from irregularities is even or level. In "Nuttall's Standard Dictionary," 1892 Edition, page 409, the following is the definition of level— "Horizontal, even, flat, on the same line of plane." This shows that level is the same as horizontal or flat, and could not possibly apply to a convex surface. In the "Cruise of the *Falcon*," by E. F. Knight, the following occurs on page 2 of volume 2:

"In the way, the rails being carried across the level plains." one perfectly straight line across the level plains."

Level here means flat or Horizontal, as the plains in South America are known to be for thousands of square miles.

"Robinson's New Navigation and Surveying," page 25, says:

"The spirit level, which is usually on the under side of the Surveyor's transit instrument, is used to determine a horizontal line. A horizontal line is at right angles lo a vertical. It is a level line."

The following is from the same work, page 33:

"To adjust a theodolite, measure very carefully the distance between two stations, and set the instrument half way between them. Now bring the level near to one of the stations, level it carefully and sight the rod. Note the number on the rod, say six feet, and have the rod man go to the other station and place his target on the rod just six feet. When the telescope is turned upon it the horizontal spider line ought to just coincide with the target, and will, if the instrument is level or in perfect adjustment."

From the foregoing it is very clear that level means horizontal and cannot mean convex.

G. F. Chambers, in his "Story of the Solar System," pages 84 and 85, quotes Sir H. Holland as seeing the eclipsed moon with the sun above the horizon. I quote the following from Mr. Chambers:

"This spectacle requires, however, a combination of circumstances rarely occurring—a perfectly clear eastern and western horizon, and an entirely level intervening surface such as that of "the sea or the African desert."

In a lunar eclipse such as described, the sun is distant from the moon half a circle, or 180, both luminaries being go from the observer, so that on a convex surface it would be impossible to see both bodies at the same time, but quite possible from a level or horizontal surface, which actually was the case. To see about 6,000 miles to the sun on the one side and about 6,000 miles to the moon on the other side, one would require to be projected 4,000 miles into space above the horizon of the globe in order to overcome the convexity in the distance. Thus, level, we are again assured, means horizontal or flat, or on the same line of plane, as the dictionary informs us. "In the "Voyage of a Naturalist," by C. Darwin, page 328, the following is stated:

"I was reminded of the Pampas of Buenos Ayres by seeing the disc of the rising sun intersected by an horizon level as that of the ocean."

The surface here referred to was a flat one, and such are called Llanos or level fields in South America. Level, therefore, signifies flat or horizontal.

Eighteen - Lighthouses

The distance at which lights can be seen at sea entirely disposes of the idea that we are living on a huge ball.

From a tract, "The Bible *versus* Science," by J. C. Akester, Hull, I extract the following:

"A lighthouse on the Isle of Wight, 180 feet high (St. Catherine's), has recently been fitted with an electric light of such penetrating power (7,000,000 candles) that it can be seen 42 miles. At that distance, according to modern science, the vessel would be 996 feet below the horizon."

Extract from a letter written by a passenger on board the "Iberia Orient Line, R.M.S. — "At noon on Thursday, 27th of September, we were 169 miles from Port Said; by the ship's log, our rate of steaming was 324 miles in 24 hours. At 12 p.m., we were alongside the lighthouse at Port Said, it having become visible at 7.30 when it was about 58 miles away. It is an ordinary tower, about as high as the tower at Springhead (60 feet), lit by electricity." According to modern science, the vessel would be 2,182 feet below the horizon.

Extract from "Manx Sun," July 24th, 1894. — "The weather of late has been very fine. It was a splendid sight, on Sunday evening, to see the land in Ayr, and Cumberland, so clear that houses could be seen with the naked eye; and the smoke from Whitehaven, and other towns, could be seen very distinctly. Ramsey Bay appeared as if it was enclosed by the surrounding laud, from Black Coombe to the Point of Ayr, Welney light being seen distinctly, distance 45 miles."

In February, 1894, a discussion on the subject of the shape of the world was carried on in the columns of the *Cape Argus* (Capetown), by the writer on the one side, and three antagonists on the other. From the evidence of the editor of the paper in a foot-note to the first letter of "Ancient Mariner" that Dassen Island light had been seen from the beach road at Sea Point, it was shewn that the water is level. This light is 155 feet above sea level at its focal plane, and according to the published report of the Inspector of Public Works for 1893, had been seen from the bridge of a mail steamer more than 40 miles away. This "ancient mariner" did not believe, and asked "if anything had gone wrong with the shape of thc earth hereabouts." One of his supporters, in a letter to the paper—after the editor had stated that the light had been seen from the beach road at Sea Point (33 miles)—stated that by climbing a hill so many feet the light might be seen! Thus will ignorant prejudice flaunt itself in the face of truth. If the earth were a globe it is evident that Dassen Island light could not be seen from a steamer's bridge 40 miles away, nor from an elevation of 30 feet at a distance of 33 miles. In the former case, allowing 40 feet for altitude of observer, the light would be 871 feet below the horizon, and in the latter 551 feet below. At the close of the controversy, I challenged "Ancient Mariner" to test the case by an appeal to an experiment on the waters of Table Bay, and am still waiting an acceptance of that challenge. I am now credibly informed that the Bluff light, Natal, has been seen at sea from a distance of 30 miles. This light is 282 feet above sea level, and should, according to the globe theory, have been 298 feet below the horizon, allowing 20 feet for height of observer!

Another and an unconscious witness to the fact of the horizontality of water, is Mr. Smith, of Cape Point, as the following shows:

A LIGHT FROM AFAR.

To the Editor of the "Cape Times."

Sir,—At nine o'clock this evening the Danger Point light was distinctly visible to the naked eye from the homestead at Cape Point (about 150 feet above sea level), this being the first occasion, since the erection of the Danger Point Lighthouse, on which the flashes of light have been noticed by myself. The light must be most powerful to be seen from a distance of over fifty miles on a clear night. I timed half a minute interval between each three quick flashes. I am, &c.,

Cape Point, August 22nd, 1894. A. E. SMITH.

In a letter from the Engineer of Public Works, dated Capetown, 2nd February, 1898, I am informed that

"The focal plane of Point Danger Lighthouse is elevated 150 feet above high water level."

According to this, therefore, if the world be a globe, the light should have been 1,666 feet below Mr. Smith's line of sight.

In Answers of 2nd May, 1896, the following appears:

"The steeple, or stump, as it is locally called, of the Parish Church of St. Botolph, at Boston, on the southeast coast of Lincolnshire, near the Wash, has long been utilised as a lighthouse. The tower is 290 feet in height, and resembles that of Antwerp Cathedral, being crowned by a beautiful octagonal lantern. This tower BEING VISIBLE 40 MILES DISTANCE serves as a lighthouse to guide mariners when entering what are called the Boston and Lynn Deeps."

According to globular principles this light should be hidden below the horizon for nearly 800 feet.

From "Music and Morals," by H. R, Haweiss, I extract the following:

"The Antwerp spire is 403 feet high from the foot of the tower; Strasburg measures 468 feet from the level of the sea, but less than 403 feet from the level of the plain. By the clear morning light, from the steeple at Notre Dame at Antwerp, the panorama can hardly be surpassed; 126 steeples may be counted, tar and near. Facing northward the Scheldt winds away until it loses itself in a white line, which is none other than the North Sea. By the aid of a telescope ships can be distinguished out on the horizon, and the captains declare they can see the lofty spire at ONE HUNDRED AND FIFTY MILES distant; Middelburg at 75 miles, Fleesing 65 miles, are also visible from the steeple; looking towards Holland, we can distinguish Breda and Wailadue, each about 54 miles off."

The above spire would be out of sight A MILE BELOW THE HORIZON, at a distance of 150 mites, and as no telescope can piece a segment of water, the conclusion is that water is level.

The *Earth Review* of July, 1894, says:

"The Captain of the S.S. *Milo,* referring to the question as to how far a powerful light can be seen, says: 'The other day, when off Skagen, the rays from Hantsholmen lighthouse were distinctly visible, though the light was fully seventy-two miles away.'"

"Mr. B. wrote and asked how the light could be seen unless the lighthouse was 3,500 feet above sea-level? This is the official reply he received.

"Editorial Department, Tit-Bits,"

"Dec. 21, 1892.

"The paragraph you refer to was sent me by the Captain of the S.S. *Milo,* and he vouched for its accuracy. Under these circumstances I cannot enter into a discussion as to the possibility of his being able to see it or not, P.S.—Mr. B, allowed that the reported observation was made from a mast-be 100 feet above sea-level."

In the *Argus Annual* for 1894, it is stated, on pages 207 and 271:

"Natal Bluff light, 292 feet above water level, has been seen at a distance of 30 miles."

According to globe measurement it should have been about 300 feet below the tine of sight.

The *Natal Mercury* of 18th July, 1898, states;

"The Cape L'Agulhas lighthouse is to be reconstructed to allow of the introduction of a flash light. A lighthouse erected two miles from Fish River, has been completed. The tower is 33 feet high and 238 feet above sea level, and *the flash light is visible for over 50 miles.*"

This light would be 1,400 feet below an observer's line of sight at an elevation of 28 feet, if the world is a globe. The following is extracted from *Scraps* of 27th August, 1898:

"I have recently received the following letter, which, I confess, fogs me just about as much as the writer of it complains of being fogged:

"Sir, — In your issue No. 772 you give an account of the lighthouse of New York — "Liberty enlightening the World." You say the light can be seen sixty miles away at sea, and I think you must be mistaken. A text-book I have by me on surveying and levelling gives eight inches per mile (actually 7.962 inches) as the correction to be made for curvature of the earth's surface in setting out canals, railways, &c., varying inversely with the square of the distance, thus; 60 x 60 x 8 ÷ 12 = 2,400 feet, and making allowance for the light being 336 feet above sea level, it should be 2,074 feet below the horizon at sixty miles.'

"'Now (1) either your figures are wrong, or (2) the weight of the statue has flattened the earth for sixty miles round about, or (3) surveyors do not allow eight inches for curvature, and let their canals and railways stick out over the side of the earth like gigantic fishing-rods. I confess I am in a fog. Can you enlighten me in your "Facts and Fancies" column? — Yours truly,

'Foggy.'

"I won't attempt to analyze "Foggy's" fogging calculations, but he is certainly very wrong. Any navigator will tell you that file horizon is visible at about fifteen miles from the hurricane deck of a steamer; at twenty from the bridge deck; and at a proportionately greater distance from the masthead. But beyond this you have to remember the added penetration given to lighthouse lights by means of refraction and reflection."

A light can only be seen on the surface of a globe, at a distance the square of which multiplied by 8" (nearly) is equal to its height. This applies no matter how powerful the light may be, because no light can pierce deep water, nor can the natural eye with or without the glass do so.

But, says someone, there is no allowance made for refraction in any of the foregoing calculations. That is quite true, but constitutes no valid objection in the light of the following extract from the "Encyclopaedia Britannica," article "Levelling":

"We suppose the visual rays to be a straight line, whereas on account of the unequal densities of the air at different distances from the earth, the rays of light are incurvated by refraction. The effect of this is to lessen the difference between the true and apparent levels, but in such an extremely variable and uncertain manner that if any constant or fixed allowance is made for it in formula or tables, it will often lead to a greater error than what it was intended to obviate. For, though the refraction may at a mean compensate for about one-seventh of the curvature of the earth, it sometimes exceeds one-fifth, and at other times does not amount to one-fifteenth. We have, therefore, made no allowance for refraction in the foregoing formulae."

We are fairly entitled to conclude, therefore, from the reliable data furnished as to how far lights at sea can be seen, that the world is an extended plane, and not the globe of astronomical speculation.

Nineteen - The Midnight Sun

M. Paul B. du Chaillu, published, a few years ago, a work entitled "The Land of the Midnight Sun," of which the following are extracts:

"The sun at midnight is *always north of the observer,* on account of the position of the earth. It seems to travel AROUND IN A CIRCLE, requiring twenty-four hours for its completion, it being noon when it reaches the greatest elevation, and midnight at the lowest. Its ascent and descent are so imperceptible at the pole, and the variations so slight, that it sinks south very slowly, and its disappearance below the horizon is almost immediately followed by its reappearance."

"We have here spoken as if the observer were on a level with the horizon; but should he climb a mountain, *the sun of course will appear higher*; and should he, instead of travelling fifteen miles north, climb about 220 feet above the *sea level* each day, he would see it the same as if he had gone north; consequently if he stood at the arctic circle at that elevation, and had an unobstructed view of the horizon, he would see the sun one day sooner. Hence tourists from Haparanda prefer going to Avasaxa, a hill 680 feet above the sea, from which, though eight or ten miles south of the Arctic Circle, they can see the midnight sun for three days,"

"As the voyage drew to a close, and we approached the upper end of the Gulf of Bothnia the twilight had disappeared, and between the setting and rising of the sun hardly one hour elapsed."

"Haparanda is in 65° 51' N. lat., and forty-one miles south of the Arctic Circle. It is 1° 18' farther north than Archangel, and in the same latitude as the most northern part of Iceland. The sun rises on the 21st of June at 12.01 a.m., and sets at 11.37 p.m. From the 22nd to the 25th of June the traveller may enjoy the sight of the midnight sun from Avasaxa, a hill six hundred and eighty feet high, and about forty-five miles distant, on the other side of the stream; and should he be a few days later, by driving north on the high road he may still have the opportunity of seeing it."

If the earth be a globe, at midnight the eye would have to penetrate thousands of miles of land and water even at 65° North latitude, in order to see the sun at midnight, That the sun can be seen for days together in the Far North during the Northern summer, proves that there is something very seriously wrong with the globular hypothesis. Besides this, how is it that the midnight sun is never seen in the south during the southern summer? Cook penetrated as far South as 71°, Weddell in 1893 reached as far as 74°, and Sir James C. Ross in 1841 and 1842 reached the 78th parallel, but I am not aware that any of these navigators have left it on record that the sun was seen at midnight in the south.

Captain Woodside of the American barkentine *Echo,* at Capetown on 26th June, 1898, reports that he had been a good deal in the great southern ocean, and often when in latitude 62° south he has had a kind of daylight all night, but not sufficient to read by; but the midnight sun was never seen.

Since writing the foregoing I have received from the Secretary of the Royal Belgian Geographical Society a paper, entitled EXPEDITION ANTARCTIQUE BELGE.

In this paper it is stated by Lieut, de Gerlache, the Commander of the expedition, that

"On 17th May the sun set, and was not seen above our horizon again until 21st July."

This was during the severest part of the winter at latitude 71° 36' south.

On pages 9 and 10 of the same pamphlet it is stated I that the ship quitted her winter quarters on the 14th February. She had thus been a winter and a summer in the ice at that latitude. During the winter, the extraordinary phenomenon of total darkness caused by the total disappearance of the sun for two months is duly recorded, and had the sun been seen at midnight in the summer, it is only natural and reasonable that such another extraordinary phenomenon should have been chronicled; but there is not one word in the pamphlet about the matter. We conclude, therefore, that there is no midnight sun in the south. The midnight sun can be seen in the north during the summer at 66° of latitude, and if there be the same extraordinary phenomenon in the south, it must have been seen at the latitude the "Belgica" reached much sooner and longer than it is in the north at latitude 66.

Twenty - Motions of the Earth

In "The Story of the Heavens," by Sir R. Ball, the following accounts of the motions of the earth-globe are given, page 3:

"It became certain that whatever were the shape of the earth, it was at all events something detached from all other bodies and poised *without visible support* IN SPACE."

Page 6:

"Ptolemy saw how this mighty globe was poised in what he believed to be the centre of the universe."

Page 7:

"Copernicus PROVED that the appearances presented in the daily rising and setting of the sun and stars *could be accounted for* by the SUPPOSITION that the earth rotated."

"The second great principle which has conferred immortal glory on Copernicus, assigned to the earth its true position in the universe. Copernicus *transferred the centre* to the sun, and he established the somewhat humiliating truth that our earth is merely a planet."

Page 87:

"The *discovery* that our earth *must* be a globe isolated in space, WAS IN ITSELF A MIGHTY EXERTION OF HUMAN INTELLECT."

Page 517:

"We *know* that the earth rotates on its axis once every day."

After all this unsound speculation, of which we know every line to be false, it is somewhat amusing to listen to another "Professor" of equal authority with the Astronomer Royal of Ireland.

Professor J. Norman Lockyer, in his "Astronomy," section IV.,

"You have to *take it as proved* that the earth moves. Day and night are *the best proofs* that the earth does really spin. Without this spinning there could be no day and night, so that the regular succession of day and night is caused by this spinning. Hence the appearances connected with the rising and setting of the sun *may be due, either to our earth being AT REST* and the sun and stars travelling round it, *or the earth itself turning round,* while the sun and stars are at rest."

"Our earth" seems to give more trouble to astronomers "than all the heavenly bodies put together. If, as Professor Lockyer says, EITHER THE EARTH IS AT REST and the stars moving, or *the stars at rest* and the earth moving, how is it that the wise men of the observatories have never once attempted to ascertain data to prove whether it is *the earth* or *the stars* that move? How is it that they are content to go on year after year, labouring under what is at best but a *supposition* that the earth moves, WHEN THE PHENOMENA. ACCORDING TO THEIR OWN SHOWING, MAY BE AS WELL ACCOUNTED FOR either by the earth being at rest, and the sun and stars moving, or the sun and stars being at rest and the earth moving.'

In "Wonders of the Sun, Moon, and Stars," by Russell, it is stated that:

"The speed of the surface of the earth, in performing its rotations, is 1,526 feet per second. Great as that speed is, it is slow when compared to the earth's progress in its orbit, which is at the rate of 18 miles per second, or *more than* 65,000 *miles per hour.*"

Then, in "The Story of the Heavens," page 429, are informed by Sir R. Ball, that:

"Every half hour we are about 10,000 miles nearer to the constellation of Lyra...the sun and his system must travel at the present rate for more than *a million years* before we have crossed the abyss between our present position and the frontiers of Lyra."

"Sun, Moon and Stars," by A. Giberne, states that:

"It is the earth that moves, and not the sun; it is the earth that moves, and not the stars."

From these extracts the reader is given to side by those who have made astronomy their life study, and, therefore, *ought* to know, that IN ONE HOUR:

"The earth rotates over 1,000 miles, revolves around the sun, over 65,000 miles, and rushes through space towards the constellation Lyra, a distance of 20,000 miles."

The total rate of rotation, revolution and gyration, amounting to no less than 50,000 miles an hour.

This casts a total eclipse over all that Jules Verne ever wrote. Put together all the imaginary exploits in the air specially written to interest the young, add to this all the wonderful adventures of airships recorded in the "Daughter of the Revolution," and tack on to this all the wild and impossible things found in "current libraries of fiction," and I venture to say that the grand total will record nothing so utterly impossible or so supremely ridiculous as this modern scientific delusion of a globe spinning away in space in several different directions at the same time, at rates of speed which no man is able to grasp; with the inhabitants, some hanging heads down and others at various angles to suit the inclination.

Write down all the swindles that ever were perpetrated; name all the hoaxes you ever heard of or read about: include all the impostures and bubbles ever exposed; make a list of all the snares that popular credulity could ever be exposed to, and you will fail in getting within sight or hearing of an imposture so gross, a hoax so ingenious, or a bubble of such gigantic projections as has been perpetrated and forced upon unthinking multitudes in the name of science, and as proved incontrovertible fact, by the expounders of modern astronomy.

Again and again have their theories been combated and exposed, but as often have the majority, who do not think for themselves, accepted the popular thing. No less an authority in his time than the celebrated Danish astronomer, Tycho Brahe, argued that if the earth revolves in an orbit round the sun, the change in the relative position of the stars thus necessarily occasioned, could not fail to be noticed. In the "History of the Conflict between Religion and Science," by Dr. Draper, pages 175 and 176, the matter is referred to in the following words:

"Among the arguments brought forward against the Copernican system at the time of its promulgation, was one by the great Danish astronomer, Tycho Brahe, originally urged by Aristarchus against the Pythagorean system, to the effect that, if, as was alleged, the earth moves round the sun, *there ought to be a change in the relative position of the stars*; they should seem to separate as we approach them, or to close together as we recede from them...At that time the sun's distance was greatly underestimated. Had it been known, as it is now, that the distance exceeds 90 million miles, or that the diameter of the orbit is more than 180 million, *that argument would doubtless have had very great weight.* In reply to Tycho, it was said that, since the parallax of a body diminishes as its distance increases, a star may be so far off that its parallax may be imperceptible. THIS ANSWER PROVED TO BE CORRECT."

To the uninitiated, the words "this answer proved to be correct," might seem to settle the matter, and while it must be admitted that parallax is diminished or increased according as the star is distant or near, *parallax* and *direction* are very different terms and convey quite different meanings. Tycho stated that the *direction* of the stars would be altered; his critics replied that the *distance* gave no sensible difference or *parallax.* This maybe set down as ingenious, but it is no answer to the proposition, which has remained unanswered to this hour, and is unanswerable.

If the earth is at a given point in space on say January 1st, and according to present-day science, at a distance of 190,000,000 miles from that point six months afterwards, it follows that the *relative position* and *direction* of the stars will have greatly changed, however small the angle of parallax maybe. THAT THIS GREAT CHANGE IS NOWHERE APPARENT, AND HAS NEVER BEEN OBSERVED, *incontestably proves that the earth is at rest* —that it does not "move in an orbit round the sun."

That the earth does not "rotate on its axis" is proved by the fact that no observer on the surface of a globe could see half way round it, or for a distance of thousands of miles on either side of him, as he would require to do in order to see round a circle of 180°, to view the setting sun and the rising moon at one time.

Sir Henry Holland, in his "Recollections of Past Life," says that:

"On 20th April, 1837, the moon rose eclipsed before the sun had set."

Now, on a globe of 25,000 statute miles equatorial circumference one has to be 24 feet above sea level to get a horizon of six miles, the "curvature" being 8" to the mile and varying inversely with the square of the distance.

We are thus taught to believe that what appears at all times of the day to be half a circle, or about 180°, is in reality only a few miles, as the earth rotates against the sun and thus deceives us. But the phenomenon of a lunar eclipse requires, according to astronomical doctrine, that the Earth shall be exactly midway between sun and moon, to shut off the light of the sun and thus to darken the moon. These two "bodies" being then, according to the astronomer, opposite each other and the earth between, must each be 90°, or a quarter of a circle distant from an observer on the earth's surface — that is, half a circle from one to the other. So that what astronomy, on the one hand, teaches is only a few miles distant, the horizon, is thus seen to be, according to its own showing, half a circle, for the sun is at one side of one quadrant, and the moon at the other side of another. If, therefore, the observer be on the equator when the phenomenon occurs, he can see, according to astronomical measurement, over 6,000 miles on either side of him, east and west. If in north or south latitude, he would see correspondingly less, but thousands of miles in every case. But, on the other hand, according to the popular theory, he would have to be hoisted 4,000 miles away in space for such a thing to be possible. The fact of lunar eclipses having been observed when sun and moon were both above the horizon at the time of the eclipse, and thus that the observer pierced, with the unaided eye, a distance of thousands of miles on either side of him — about half a circle — proves that the earth does not rotate, and that it is not the globe of popular belief.

Sir Henry Holland further informs us that:

"This spectacle requires, however, a combination of circumstances rarely occurring—a perfectly clear eastern and western horizon, and an entirety *level intervening surface, such as that of the sea,* or the African desert."

It is this LEVEL INTERVENING SURFACE that defies all astronomical attempts to make it convex, and proves beyond the possibility of a doubt that the earth is an extended plane and not a globe.

Furthermore, if the earth-globe rotates on its axis at the terrific rate of 1,000 miles an hour, such an immense mass would of necessity cause a tremendous rush of wind in the space it occupied. The wind would go all one way, and anything like clouds which got "within the sphere of influence" of the rotating sphere, would have to go the same way. The fact that the earth is at rest is proved by kite flying. The following from the "American Exporter" of November, 1897, illustrates this:

"Recently, a very interesting experiment was made in high kite flying at Boston, from the Blue Hill Observatory, when the highest altitude ever reached by a kite was obtained. The top kite reached a height of 10,016 feet above sea level, or 8,386 feet above the summit of the hill. At the highest point reached the temperature was 38°, while at the ground it was 63°... Above 5,000 feet the wind was from the west, while at the ground there was a southerly wind."

Astronomers are not agreed about the "depth" of the Earth's atmosphere, but the lowest estimate is 45 miles. Therefore, everything within the atmosphere would be subject to the gale of wind produced by the mad whirligig of the rotating globe. When, however, we know that "above 5,000 feet the wind was from the west, while at the ground there was a southerly wind," the fact of the earth being at rest again dawns on us. How could there be two different directions of the wind at a distance of only 5,000 feet apart, if globular hypotheses are anywhere near the truth? Spin a top and it will be seen that the rotation of the top causes the air within its sphere of rotation to go all one way.

Let "imagination" picture to the mind what force air would have which was set in motion by a spherical body of 8,000 miles diameter, which in one hour was spinning round 1,000 miles, rushing through space 65,000 miles, and gyrating across the heavens 20,000 miles. Then let "conjecture" endeavour to discover whether the inhabitants on such a globe could keep their hair on. Talk about Jules Verne, he is not in it with the expounders of this "most exact of all the sciences."

A. E. Skellam says:

"The following experiment has been tried many times, and the reasonable deductions from it are entirely against any theory of motion: A loaded cannon was set vertical by plumb-line and spirit-level and fired. The average time the ball was in the air was 28 seconds, On several occasions the ball returned to the mouth of the cannon, and never fell more than 2 feet from its base, as shown in Fig. 1 (figures omitted). Now, let us see what the result would be if the earth were a rapidly rotating sphere. The ball would partake of two motions, the one from the cannon, vertical, and the other from the earth, from west to cast, and would arrive at B, Fig. 2; while it had been ascending, the earth, with the cannon, would have gone on to C, In descending it would have no impulse from the earth's motion or from the cannon, and would fall in a straight line at C. but during the time it were falling, the earth, with the cannon, would have travelled on to D, and the ball would fall (allowing the world's rotation to be 600 miles per hour in England) more than two miles behind the cannon."

Twenty-One - The Moon

According to current science the moon was once a piece of molten rock fractured off from the earth, when the earth was in a soft or plastic condition. Its origin is thus stated by Sir R. Ball, in the "Story of the Heavens," page 520:

"There is the gravest reason to believe that the moon was at some very early period, fractured off from the earth, when the earth was in a soft or plastic condition...At this epoch the earth rotated 29 times on its axis, while the moon completed one circuit...but whether it (the epoch) is to be reckoned in hundreds of thousands of years, in millions of years, or in tens of millions of years, *must be left in great degree to conjecture*."

Conjecture, in this case, has to choose between hundreds of thousands and tens of millions of years. Ample scope one must admit! In the same volume, page 52, the insignificant size of the moon as compared to the stars is set forth:

"Every one of the thousands of stars that can be seen with the unaided eye, is enormously larger than our satellite."

In "Wonders of the Sun, Moon, and Stars," the same idea is announced thus:

"The luminary which appears to us next in importance to the sun is the moon, and for practical purposes, it, of course, is so; but, considered from a broad astronomical point of view, *the moon is exceedingly insignificant being the smallest of all the luminaries visible to us with the naked eye.* The diameter of the moon is only 2,160 miles."

The moon is said to be a reflector of the sun's light, and to have no light of her own, as the following shows. Sir R. Ball, in his "Story of the Heavens," pages 50 and 56, says:

"The brilliancy of the moon arises solely from the light of the sun which falls on the not self-luminous substance of the moon."

"The sunlight will thus pass over the earth to the moon, and the moon will be illuminated."

The speculation regarding the origin of the "lesser light that rules the night" is in keeping with the other impossible notion concerning the earth being shot off from the sun in remote ages. It is so purely nonsensical that it may well be relegated to oblivion without further ado.

As to size, the moon is next in importance to the sun, if, indeed, she is not quite as large; and many times larger than any star in the heavens, including al] the planets ever seen by the eye of man.

Both the distance and size of most of the objects in the heavens may be measured with a high degree of accuracy. It only requires to be known that the object is vertical to a certain part of the world at a certain time, when the observer must take a position—which could be ascertained by previous experiment—where the angular distance of the object is 45°. A base line measured from that position to the point at which the object was vertical at the moment of observation, will be the same length as the distance of the object from the earth's surface.

Size, except in the case of very small stars, may be as easily determined. Let the instrument with which the angular distance was taken be graduated to degrees, minutes and seconds, the minutes and seconds corresponding to miles, and sixtieths of miles on the earth's surface.

Having carefully adjusted the instrument, bring the image of the lower limb of the object to be measured down to the horizon, and note the reading on the instrument. Now bring the upper limb in contact with the horizon, and the difference of the reading will be the diameter of the object. It would, of cour.se, require a very finely adjusted instrument, and one graduated to say the one hundredth part of a second to measure some of the smaller stars.

Instead of the diameter of the moon being 2,160 miles, as we are informed by the men of science of to-day, it is, by i the above process, found to be about 32 nautical miles in diameter.

Then as to the moon being a non-luminous body, and receiving alt its light from the sun, astronomy is as hopelessly wrong as in most other of its fanciful statements.

If the reader has taken notice of reflectors, he will have seen that they are either flat — where angles are involved — or conclave, but never convex. A convex surface cannot concentrate and reflect light. But a concave surface does this, hence all reflectors, where angles are not involved, are concave. The moon is a globe. It is convex, and therefore cannot reflect light to any extent.

Then, if the moon could reflect the light, it would also reflect the heat of the sun. But we know that moonlight is cold instead of warm. In Noad's "Lectures on Chemistry," it is said:

"The light of the moon, though concentrated by the powerful burning glass, is incapable of raising the temperature of the most delicate thermometer."

"The Lancet" says:

"The moon's rays, when concentrated, actually reduce the temperature upon a thermometer more than 8°."

When light and heat are received by a reflector, light and heat are reflected, as the reader may prove for himself, by testing the matter with a petroleum lamp and a reflector.

If a red light be projected on to the surface of a reflector the reflection of it is red. In line, reflectors reflect just what they receive.

If fish be hung up to dry in the sun, they will be preserved. But if exposed to the moon, will be rendered putrid in one night. The same applies to fruits, See, clearly proving that the light of the moon cannot be of the same nature as that of the sun. And, furthermore, that the moon shines by its own light. The nearest approach to moonlight is phosphorescent light. And if the moon and stars be observed through a telescope, it will be noticed that .starlight and moonlight, except in a few cases, are identical; the .size of the star determining its brilliancy, on the principle that the larger the star the greater will its brilliancy be. "Sun, Moon, and Stars," page 57, says:

"That soft, silvery light, *so unlike sunlight,* or gaslight, or any other kind of light seen upon the earth."

The theory that moonlight is only reflected sunlight requires that the illuminated part of the moon be always next the sun. Unfortunately for the theory, however, this is not the case.

If the Moon be observed from last quarter to new, it will be found that for a portion of one day, immediately before new moon, the dark part of the moon is turned towards the sun; and at new moon the sun is still to the eastward of the moon, which is illuminated on its western surface.

On 10th August, 1898, at Durban, Natal, the moon rose at 1.7 a.m., and by sunrise (6.35) was high in the heavens, showing about half on her eastern surface. On 15th, moon rose 4.56 a.m. (sunrise 6.30), with a very small portion of eastern limb illuminated, but the whole circle was distinctly visible. On 16th, moon rose 5.32 a.m. (sunrise 6.29) with the dark part towards the sun. On 17th, moon rose 6.4 a.m. (sunrise 6.28), 24 minutes before the sun, New moon same day 6.35 p.m., the moon's illuminated western limb be turned away from the sun, which was to the eastward. On 18th, moon rose 6.36 a.m. (sunrise 6.27), and the sun was thus ahead of the moon, and on the illuminated side, having passed her between the hours of sunset on the 17th and sunrise on the 18th. Anyone who cares to take the time and make the necessary observations, may satisfy himself on this point. The almanac shows that at every new moon, the sun is to the east of the moon, which is illuminated on her western surface, clearly proving that the moon is a self-luminous body, and not a reflector of sunlight.

But how about the "phases" of the moon, if she is self-luminous? If the moon be observed it will be apparent that she rotates from west to east in order to produce the various phases, each phase appearing in spite of the position of the sun. This shows that she is luminous on half her surface, the dark half being towards us when she is invisible.

Take a wooden ball and rub half its surface with a solution of phosphorous in olive oil. Place the ball in a dark room, and cause it to rotate, and all the phases, representing those of the moon will be manifested.

It is said that the moon has been photographed and that extinct volcanoes, dry watercourses, &c., have been found on its surface. The place were seas once were, it is alleged, have not only been photographed, but named, and thus there is nothing wanting to show that the moon was once inhabited —a world like ours.

We know that "poets are licensed to lie," but astronomers who claim that their science is the most exact of any, and admits of demonstration, should be careful to speak the truth, surely. How then are photos of the moon obtained? Sir R. Ball shall tell us. In "The Story of the Heavens," note, p. 62, says:

> "This sketch has been copied by permission from the very beautiful view in Messrs. Nasmyth and Carpenter's book...So have also the other illustrations of lunar scenery in Plates 7, 8, 9. *The photographs were obtained by Mr. Nasmyth from models carefully constructed by him to illustrate the features of the moon."*

In the text. Sir Robert very carefully says that

> "This is no doubt a somewhat imaginary sketch."

Read also the following from "Answers to Questions," by W. Bathgate, M,A.:

> "The author of a work called 'The Plurality of Worlds,' says: 'Take the appearance of the heavenly bodies, the moon; examine its appearance by the best constructed telescope; read all that has been written upon it by the most skilful astronomers, and nothing remains to satisfy a mind that thinks and reasons for itself, a mind not warped by theory and fanciful hypothesis...The mountains and valleys, the seas and rivers, the fields and orchards, are all in the head of the observer. Ever since I looked at the moon through a good telescope, I have been much surprised at the credulity of the human mind in the combination of opinions raised from the appearance of this planet...These discoveries are hypothetical. You will not elicit them by applying the rules of the Baconian philosophy, or by looking through a telescope, aided by the science of geometry; BUT THEY ARE INVENTED IN THE CLOSET, BROUGHT TO THE TELESCOPE, AND THEN USHERED INTO THE WORLD AS THE CLOSE RESULT OF INDUCTIVE INVESTIGATION.'"

No, gentle reader, there are no "extinct volcanoes" on the moon; there are no "seas" on her surface. You have been badly "had" by the profession, that is all. Let photography be questioned as to the possibility of securing a correct picture of an object at a distance of 240,000 miles!

Twenty-Two - Eclipses of the Moon

From "Wonders of the Sun, Moon, and Stars" I extract the following:

"Astronomers, by mere calculation, are able to forecast the position of any luminary at any time for many years to come. By the same means, they can foretell to a second, the commencement, duration, precise aspect, and the ending of all the eclipses that will occur for a lifetime hence, and more, without limitation. *Such being the case, the theories upon which the calculations are based must be true, or the correctness of such calculations would be impossible.*"

This statement, and similar ones so often made, have had the effect desired by their inventors. The public have believed that the theory of a globular world is true, because astronomers can correctly foretell eclipses. This is a totally erroneous view of the matter, as eclipses have no connection with the shape of the world, and are not calculated on any theory, but on well-known time cycles. In "Pagan Astronomy," by A. McInnes, the following occurs:

"More than 2,000 years ago the Chaldeans presented to Alexander the Great at Babylon, tables of eclipses for 1,993 years; and the ancient Greeks made use of the cycle of 18 years, 11 days, the interval between two consecutive eclipses of the same dimensions. The last total eclipse of the sun occurred on Jan. 22 1879, and the preceding one on Jan. 11, 1861. Now, have not mere theorising about the sun and moon — the great unerring clocks of time — thrown chronology and the calendar into confusion, and hence scientists cannot agree as to the world's age, and the year absurdly begins on Jan. 1 instead of at the vernal equinox, the months consisting of 31 or 30 days, one of 28? However, it can be shown that, with ellipse and star transit cycles, the greatest accuracy as to dates may be attained.

"Going back, for example, from Jan. 11, 1861. through a period of thirty-six eclipses, or 651 years, we find that a total eclipse occurred also on Jan. 11, 1210; and, continuing backwards, by such cycles we arrive precisely at the date of creation as given by Moses in Genesis. Also, as related by Josephus, the moon was eclipsed in the fifth month of 3,998 a.m., when Herod the Great died, and Christ being then two years old, His birth occurred 3,996."

In "The Triumph of Philosophy," Mr, J, Gillespie informs us as follows:

"I am asked to take into consideration how they, with the present theory, can calculate and foretell eclipses and other events with surprising accuracy. Now, I can prove that long before the present theory was ever thought of, even 600 years before Christ, the ancients discovered the difference of local time or the hour of the day between places of different longitudes, knew the causes and laws of eclipses, and the motion of the sun, moon and stars with surprising accuracy,"

R. J. Morrison, F.A.S.L., R.N., in his "New Principia," says:

"Eclipses, occultations, the positions of the planets, the motions of the fixed stars, the whole of practical navigation, the grand phenomena of the course of the sun, and the return of the comets, may all and everyone of them be as accurately, nay, more accurately, known without the farrago of mystery the mathematicians have adopted to throw dust in the eyes of the people, and to claim honours to which they have no just title...The public generally believe that the longitudes of the heavenly bodies are calculated on the principles of Newton's laws. *Nothing could be more false*."

T. G, Ferguson, in the *Earth Review*, for September 1894 says:

"No doubt some will say, 'Well, how do the astronomers foretell the eclipses so accurately.' This is done by cycles, The Chinese for thousands of years have been able to predict the various solar and lunar eclipses, and do so now, in spite of their disbelief in the theories of Newton and Copernicus. Keith says, 'The cycle of the moon is said to have been discovered by Meton, an Athenian, B.C. 433,' when, of course, the globular theory was not dreamt of.

E. Breach, in his "Fifty Scientific Facts," says:

"Sir Richard Phillips in his Million Facts, says, 'Nothing therefore can be more impertinent than the assertion of modern writers that the accuracy of astronomical predictions arises from any modern theory. Astronomy is strictly a science of observation, and far more indebted to the false theory of Astrology, than to the equally false and fanciful theory of any modern.

"We find that four or five thousand years ago, the mean motion of the Sun, Moon and Planets were known to a second, just as at present, and the moon's nodes, the latitudes of the planets, &c., were all adopted by Astrologers in preparing horoscopes for any time past or present. Ephemerides of the planets places, of eclipses, &c., have been published for above 600 years, and were at first nearly as precise as at present."

The same thing is admitted by Sir R. Ball, in his "Story of the Heavens." On page 56, he informs us:

"If we observe all the eclipses in a period of eighteen years, or nineteen years, then we can predict, with at least an approximation to the truth, all the future eclipses for many years. It is only necessary to recollect that in 6585 1/3 days after one eclipse a nearly similar eclipse follows. For instance, a beautiful eclipse of the moon occurred on the 5th of December, 1881. If we count back 6585 days from that date, or, that is, 18 years and 11 days, we come to November 24th, 1863, and a similar eclipse of the moon took place then...*It was this rule which enabled the ancient astronomers to predict the occurrence of eclipses at a time when the motions of the moon were not understood nearly so well as we now know them*."

The foregoing extracts speak for themselves, and show clearly that the statement quoted from "Wonders of the Sun, Moon, and Stars," is entirely fallacious.

This same text book states on page 110:

"When the moon gets on the side of the earth precisely opposite the sun, the interpolation of the mass of the earth causes an eclipse of the moon."

But this statement is stripped of all its glory by the fact that lunar eclipses have taken place when both sun and moon were in full view, as Sir H. Holland informs us, and which we have before referred to.

But if there is a way to wriggle out of the logical conclusion attaching to this fact, astronomers will find it, and so we are coolly informed that refraction is the cause of the moon being visible in such a case. The moon, it is said, is really below the horizon, but refraction has cast its image upwards and thus it can be seen. To square the matter, it is stated that this refraction amounts to "over 30 minutes at the horizon," Now, 30 minutes is about the diameter of the moon, and thus it is said that the refraction is over 30 minutes at the horizon, so that the phenomenon may be accounted for, and the moon, which is in full view, declared to be actually below the horizon. But this refraction is incapable of verification. Firstly, because refraction can only operate when the moon and the observer are in different densities, and it cannot be proved that such is the case. And, secondly, if such were the case, it could not be proved that refraction amounts to over 30 minutes at the horizon. A table of refraction before me gives it as nearly 35 minutes at the horizon, and only 3' at an angle of 17½°". This is so utterly impossible, that it must be rejected.

The only object of the table for the horizon seems to be to account for the phenomenon we have mentioned. But it is really too transparent, and must be cast aside as worthless and as being an endeavour to make theoretical astronomy tally with the facts. The fact that sun and moon have been seen above the horizon at a lunar eclipse, completely disproves the theory that the earth has got between the two luminaries. Refraction cannot be proved to exist, because it cannot be proved that the moon is in a greater density I than the observer. And even if we "assume" the moon to be in such greater density the amount of it is entirely uncertain, and thus the theory in its entirety must be rejected.

E. Breach; in his "Fifty Scientific Facts," says;

"It is supposed that an eclipse of the moon is caused by the earth intervening between the sun and moon. The earth in reckoned to travel 1,100 miles per minute; how long would it be passing the moon, travelling herself at 180 miles per minute? Not (our minutes. Vet the last eclipse of the moon, on February 18th, lasted 4½ hours; so it could not be the earth intervening, as both luminaries were above the horizon when the eclipse commenced, and the spots of the moon could be seen distinctly through the shadow; the moon was also seen among the stars."

This is a hard nut for Newtonians to crack, and not quite so easy of accomplishment as "cracking the crust" of their globe theory.

But the battle is not won yet. There is another bugbear to face. It is alleged by the learned that at a lunar eclipse the earth casts a shadow on the moon, by intercepting the light of the sun. The shadow, it is alleged, is circular, and as only a globe can cast a circular shadow, and as that shadow is cast by the earth, of course the earth is a globe. In fact, what better proof could any reasonable person require? "Powerful reasoning," says the dupe. Let us see. I have already cited a case where sun and moon have been seen with the moon eclipsed, and as the earth was not between, or they both could not have been seen, the shadow said to be on the moon could not possibly have been cast by the earth. But as refraction is charged with raising the moon above the horizon, when it is said to be really beneath, and the amount of refraction made to tally with what would be required to square the matter, let us see how refraction would act in regard to a shadow. Refraction can only exist where the object and the observer are in different densities. If a shilling be put in the bottom of a glass and observed there is no refraction; but as soon as water is poured into the basin, there is refraction. Refraction casts the image of the shilling UPWARDS, but a shadow always *downwards.* If a basin be taken and put near a light, so that the shadow of the edge touches the bottom of the basin, and a rod be placed on the shadow and water be poured in, *the shadow will shorten inwards* and DOWNWARDS; but if the rod is allowed to rest in the basin and water poured in, the rod will appear to be bent UPWARDS, This places the matter beyond dispute and proves that it is out of the range of possibility that the shadow said to be on the moon could be that of the earth. Herschel admitted that there are many invisible moons in the sky, and it is just one of these that eclipses the moon, being visible as it passes over her luminous surface. But even if we admit refraction, and that to the extent seemingly required to prove that when the eclipsed moon is seen above the horizon with the sun visible, the moon is in reality below the horizon, we are still confronted with a fact which entirely annihilates every theory propounded to account for the phenomenon. Taking the astronomers' own equation of 8" to the mile, varying inversely as the square of the distance, for the curvature of the earth, where sun and moon are both seen at a lunar eclipse, the centre of the sun is said to be in a straight line with the centres of the earth and the moon, each luminary being go6uj from the observer. This would give about 6,000 miles as the distance of each body from the observer. Now, what is the curvature in 6,000 miles? No less than 24,000,000 feet or 4,545 miles. Therefore, according to the astronomers own showing an observer would have to get up into space 4,545 miles before he could see both sun and moon above his horizon at a lunar eclipse!!! As lunar eclipses have been seen from the surface of the earth with sun and moon both above the horizon at the same time, it is conclusively proved THAT THERE IS NO "CURVATURE OF THE EARTH," and, therefore, that the world is a plane, and cannot by any possibility be globular. This one proof alone demolishes for I ever the fabric of astronomical imagination and popular credulity.

In The *Belfast News Letter*, there appeared the following letter:

To the Editor of the Belfast News Letter.

"Sir,—I have been requested to direct attention to the forthcoming eclipse of the moon, which will take place on the 28th instant, and have much pleasure in doing so.

"On Friday next this interesting phenomenon will take place during the ordinary observing hours of the evening, and , will, no doubt, attract some attention should the weather prove favourable. The first contact of the disc of the moon with the shadow of the earth will take place al about eight minutes to six o'clock in the evening; the middle of the eclipse happening at twenty-two minutes past seven o'clock; and the last contact of the moon's disc with the earth's shadow will take place about nine o'clock p.m. The eclipse will be a partial one, hut a large area of the lunar disc will be immersed in the shadow of the earth. If the diameter of the moon be taken as unity, the magnitude of the eclipse will be 0.87. The first contact of the lunar disc with the shadow may be looked for at 85° eastward from the northernmost portion of the limb of the moon; and the last contact with the shadow will take place at 30° from same starting point in a westerly direction.

"It will be interesting to those people who have recently been treated to a dissertation on the non-rotundity of the earth by i member of the so-called Zetetic Society (an association formed with the object of proving, amongst other things scarcely orthodox from an astronomical point of view, that the earth is not a sphere, but is rather a great flat plain, to watch the well-defined circular shadow which the earth will, by its interposition between the sun and moon, cast upon the disc of the latter body.—Yours truly,

W. REDFERN KELLY. F.R.A.S.

Dalriada, Malone Park, Belfast,
24th February, 1896."

In a subsequent issue of the paper the following appeared:

To the Editor at the Belfast News Letter.

"Sir,—Having come across Mr. W. Redfern Kelly's letter on the above in your issue of the 25th, it occurred to me that the writer is mistaken in thinking the Zetetic Planeist's (as they call themselves) ideas ran be injured or swept away by such superficial remarks. Unfortunately for the globular side, many eclipses have taken place when the sun has been above the observer's horizon, thus nullifying at once the generally accepted idea that it is the shadow of the intervening earth projected on the moon by the sun. Again, the moon is recorded to have been eclipsed by a triangular shadow. This, of course, makes the Newtonians' case still worse. As to the accepted idea that the foretelling of eclipses proved the truth of the Newtonian hypothesis, this must be only mentioned to be ignored, it being well known and allowed by those who have studied this branch of astronomy to be merely a matter of correct observations during a series of years to foretell the exact lime of either lunar or solar eclipses for an indefinite number of years, and has nothing whatever to do with the shape of the world,

"I trust the writer of the letter in question and other champions of the Newtonian system In Belfast will see the weakness of their attack in this instance, and take counsel, so as to attack these stubborn-minded globe-smashers or planeists in a more vulnerable position. Apologising for trespassing on your valuable space, and thanking you in anticipation foe inserting my letter.—I am, dear sir, yours,

March 10th. H. H. D'ARCHY ADAMS."

The following letters, published in the *Earth Review*, in 1896, were refused insertion in the *Belfast News Letter*:

To the Editor of the Belfast News Letter.

"Sir,—In your issue of yesterday, I observe an article by Mr. Redfern Kelly, relative to the coming lunar eclipse. In that article reference is made to the Zetetic Society and its contention, viz.:—that the earth is not globular. This, indeed, is the contention, and the Society is indebted to Mr. Kelly for the opportunity thus afforded of giving some of their views publicly, particularly in this instance with regard to eclipses. Now, the fact may be gainsaid, but cannot be logically deemed, that the surface of standing water is horizontal. Water has been proved repeatedly by the Zetetic School to be fiat or level, without curvature. Such being the case the earth must and does conform to that configuration with the sun and moon above the surface. With such conditions it is obvious a shadow of the earth cannot operate, both luminaries being overhead, and several instances are on record where eclipses have taken place when sun and moon have been above the horizon, the earth being out of range of both. Of course it will be argued that refraction operated in such cases, and at first this, explanation may appear plausible, but on carefully examining the subject it is found to be inadequate, and those who have recourse to H cannot be aware that the refraction of an object and that of a shadow are in opposite directions. An object by reaction is bent upwards, but the shadow of any object is bent downwards, as will be seen by the following simple experiment:—Take a plain white shallow basin, and place it ten or twelve inches from a light in such a position that the *shadow* of the *edge* of the basin touches the centre of the bottom. Hold a rod vertically over and *on* the edge of the shadow, to denote its true position; now let water be gradually poured into the basin, and the shadow will be seen to recede or *shorten inwards* and *downwards,* but if a rod or a spoon is allowed to rest, with Its upper end toward the light, and the lower end in the bottom of the vessel, it will be seen as the water is poured in to bend *upwards* - thus proving that if refraction operated at all it would do so by elevating the moon above its true position, and throwing the earth's shadow downwards, or directly away from the moon's surface. Hence it is clear that a lunar eclipse by a shadow of the earth is not possible. It is admitted by Herschel and other astronomers that invisible bodies exist in the firmament, and such an amount of evidence on this point has accumulated as to put the matter beyond all doubt — such bodies, though invisible to the naked eye, become apparent when in a line between an observer and a luminous body like the moon, the intervention of such a body (says the celebrated Zetetic Astronomer known as "Parallax") i is the direct cause of a lunar eclipse. There are instances on I record showing that some other cause existed than that of the earth's shadow to produce an eclipse.

"Mr. Walker, who observed the lunar eclipse of March 19th, 1848, near Collumpton, says, the appearances were as usual until twenty minutes past nine, at that period, and for the space of the next hour, instead of an eclipse or shadow (umbra) of the earth being the cause of the total obscurity of the moon, the whole phase of that body became very quickly and most beautifully *illuminated,* and assumed the appearance of the glowing heat of fire from the furnace, rather tinged with a deep red, the *whole disc* of the moon being as *perfect with light* as if there had been no eclipse whatever. THE MOON POSITIVELY GAVE GOOD LIGHT FROM ITS DISC DURING THE TOTAL ECLIPSE.' Of course it will be asked how the phases of the moon can be accounted for on the Zetetic basis. The reply is, the moon is semi-luminous, shining with an *independent light of its own*, one side is illuminated and the other not, as it revolves, all the phases we are familiar with become apparent. That the moon is not a perfectly opaque body, but a crystalised substance, is shown from the fact that when a few hours old or even at quarter we can through the unilluminated portion see , the light shining on the other side. Stars have also been observed through her surface. In conclusion (for I have already transgressed with regard to valuable space), I would observe that a system requiring for its support such a condition and such belief as that associated with the antipodian theory, must necessarily be absolutely theoretical, and consequently devoid of *any facts!*

26th February, 1896. J. ATKINSON."

To the Editor of the Belfast News Letter.

May I, with your kind permission, ask W. Redfern Kelly, Esq., F.R.A.S., to answer in your columns the following questions:

1st. — Prove by any practical demonstration that it is "the shadow of the earth" that eclipses the moon.

2d. — Why is it that the 'shadow' is not always a globular one, and not always the same size?

3rd. — As the duration of the eclipse of the moon on February 28th lasted 3 hours 8 minutes, will he kindly explain why eclipses in Ptolemy's time lasted over 4 hours?

4th.—Is it not possible that one of the 'dark bodies' which Anaxagoras said were lower than the moon and move between it and the earth is the cause of lunar eclipses? If not, why not?

5th.—Will he, by a *practical experiment upon the earth's surface*, or surface of standing water anywhere in the world, give us ONE proof that the earth is an oblate spheroid?

Awaiting his esteemed replies, which I trust, for the elucida of Truth, you will allow me to reply to.—I remain, yours respectfully,

J. WILLIAMS, *Hon. Secretary,*

Universal Zetetic Society,
32, Bankside, London, S.E.

To the Editor of the Belfast News Letter.

Sir,—In your issue of Tuesday, February 25th, I noticed a letter referring Zetetics to the eclipse of the moon on the 28th of the same month, for a proof of the supposed globularity of the earth.

If the writer had first given proof that it is the shadow of the earth which falls upon the moon, there would have been some support for his contention; but he, like all astronomers, first *assumes* that it is 'the shadow of the earth,' and secondly, that nothing but a globe can cast a circular shadow! Let him clear his argument, if we can call it one, of these underlying assumptions which vitiate it, by giving some proof of his premises, then I will, with your kind permission, examine whether his conclusions necessarily follow.

I, as one of those Zetetics your correspondent refers to, did watch the eclipse as far as the cloudy state of the sky would permit, and I must state that I drew conclusions from the phenomena very different horn those he would draw, and in favour of the Zetetic position.

As Mr. Kelly seems kindly disposed towards the 'so-called Zetetic Society,' and seeks to instruct them in correct astronomical principles, he will, perhaps, after giving the proofs above asked for, be good enough to instruct us on the following points:

(1) Why did the 'shadow of the earth' begin to obscure the moon's light on her eastern limit?

(2) Why did the 'shadow' not go right across the moon's disc, *i,e.* in the same general direction, as all the bodies involved continued in the same course as they were in when the eclipse commenced?

(3) Why did the 'shadow,' after commencing to obscure the moon on her left or eastern edge, gradually disappear at the top or upper surface of the moon?

(4) If the moon's light be only reflected sunlight, why is not all that light cut off when the earth is supposed to come in between the sun and the moon? In

other words, how is it the moon's disc can be dimly seen when and where the illuminating light is cut off, even to the extent of a total eclipse? And

(5) Can your correspondent give us any testimony whatever, not vitiated by astronomical hypothesis, going to prove that the earth, which ordinarily feels so stable, has any of the awful motions attributed to it?

If facts can be shown in answer to the above questions, and in favour of the popular contention, I can promise your correspondent that his efforts will not be thrown away on Zetetics, because, as far as I am acquainted with them, and as their name implies, they are honest and fearless investigators of the truth in these matters. — I am. Sir, yours respectfully,

ALBERT SMITH."

23, East Park Road, Leicester.

It is thus left on record that the columns of the *Belfast News Letter* were closed to that open and above-board discussion for which the Press should be celebrated. A "Free Press" is what is wanted, so that the public may have both sides of the matter before them and thus be able to judge for themselves. But it is mostly the other way. Letters dealing, with unpopular subjects, or taking a side against the commonly-accepted "view," are often consigned to the waste-paper basket. In this connection, however, I desire to bear witness to the freedom of the Press in this Colony. Nowhere in the world, is there that liberty and freedom of thought that should characterise a free people, as is found in Natal. At least that is my opinion. And certainly I know of no other place that can boast of such impartiality in the matter of newspaper correspondence as enjoyed by the people of Durban.

I have now finished my dissertation on the theories of astronomers regarding the moon, and we have seen that, as in every other case we have considered, there is not a word of truth in the statements of the "learned" concerning the "lesser light that rules the night."

Twenty-Three - Magnetism

In a work entitled "Magnetism and Deviation of the Compass." by J. Merrifield, L.L.D., F.R.A.S., 10th Edition, page 4, the statement is made that:

"When a magnet is suspended by a thread without torsion or on a pivot so as to move freely, it will, when left to itself, rest only in a vertical plane which stands nearly North and South."

If this statement be read with an artificial globe in sight, the assurance is at once conveyed to the mind that the shape of the world cannot be globular. On a vessel at sea, the compass needle could not point nearly north and south on a globular surface, but would point into the sky at both ends. To point north on the equator it would dip towards the North Pole at an angle of 45°, while the south end would be the same angle above the horizon, pointing into the sky. Only on a flat surface could the statement of Dr. Merrifield be true. What we know is that the compass needle is horizontal, except in high latitudes, and there, although it dips, spins round, and does various other extraordinary things, no constant of dip can be found. It is never the same at the same latitude at different times. In fact, there is nothing yet discovered that accounts for the deviation of the compass, lateral and vertical.

In an article in the *Nineteenth Century*, 1895, by C. R. Markham, it is stated that:

"Professor Neumayer writes that without an examination and survey of the magnetic properties of the Antarctic regions, it is utterly hopeless to strive with prospects of success, at the advancement of the theory of the earth's magnetism."

It is confessed that our knowledge of what is called the earth's magnetism is very scanty. The *Journal of the Society of Arts* of 20th March, 1896, says:

"Magnetical observations in the south are at present not only urgently needed for the purpose of navigation, but also for supplying a missing link in our knowledge of terrestrial magnetism."

And Lord Kelvin, speaking at Burlington House, on 30th November, 1893, stated:

"We are certainly far from having any reasonable explanation of any of the magnetic phenomena of the earth."

It is evident that the sun has something to do with magnetism, as disturbances of sun spots have often been accompanied by disturbances of magnetic needles.

The dipping needle is an instrument constructed to record the dip at various latitudes. But as this instrument does not allow of the needle moving in a lateral direction, it is useless for any determination of the deviation of the horizontal needles. It has been claimed that it proves the globular .shape of the earth, by recording the dip of the horizontal needles. This, however, it does not, and in its very construction cannot do, for the reason above stated. In London, in latitude 51½ north, the dipping needle experiments should show that the dip is that amount, if the theory be true. In "Magnetism," by Sir W. Snow Harris, page 163, it is recorded that:

"Sabine in 1821 determined the inclination in London by the two methods of oscillation and by Mayer's needle, and arrived at the three following results:- Mayer's needle, 70° 2' 9"; methods of oscillation, 70° 4' and 70° 2' 6"."

It is evident, therefore, that we have not yet sufficient information regarding magnetism to lay down any definite rules for determining the cause and cure of deviation, whether lateral or vertical. In Harris' "Magnetism," page 254, it is stated that "Our planet is a magnet," and "that a magnetic bar is horizontal at the equator, and that in north latitude the north end of the bar dips *towards the south,* while in south latitude the south end dips *towards the north.*" That is to say, in both north and south latitudes the compass points *upwards.* This is clear from the figure (127, page 254).

In "Magnetism and Electricity," by W. G. Baker, we find an illustration of the same supposed principle on page 16. Unfortunately for the exposition of Sir W. Snow Harris the figure accompanying the text states the case to be the very reverse of that gentleman's figure.

In this figure, the bar dips *down* from the centre of the magnetic field — the equator — towards both north and south.

Both these books are standard works on the subject of "Magnetism," and yet in this, the most important of all points, they are exactly opposite!

The statement of Sir W. Harris will not bear investigation. It may be an easy way "explaining" (which the learned are good at), but it does not agree with fact.

Mr. Norman H. Pollock, writing from 115, Broadway, New York, on 4th December, 1897, informs me as follows;

"Your letter of enquiry dated Nov. 2 received. I am sorry that I cannot throw much light on the subject of the 'dip' of the compass. The vessel I was on was a wooden steamer, copper-fastened. With the exception of the engine, and anchor and chains, there was no iron about her. The compass worked well until we were about 100 miles from the entrance to Hudson Straits, when they became utterly useless. We had about thirty of them, and no two pointed in the same direction. When whirled around they did not stop towards the north, but in all directions, and when they did stop the needle was depressed about 45° and usually stationary...I was on shore (nothing but rock) and did not see iron ore...The highest latitude was about 67°..."

It is well known that magnetism acts in a straight line. This of itself is sufficient to prove that the earth cannot be a globe; because on a globe, wherever the magnetic influence came from, the needle would point in that direction; sometimes down through the ship's keel, and always at an angle that would render it useless to the navigator. The truth about magnetism has yet to be discovered, but even in our present state of knowledge, the weight of evidence goes to show that the world cannot be the globe of popular belief.

Twenty-Four - Navigation

It must be obvious to the reader that, if the earth be the globe of popular belief, the rules observed for navigating a vessel from one part of this globe to another, must be in conformity to its figure. The *datum line* in navigation would be an arc of a circle, and all computations would be based on the convexity of water and worked out by spherical trigonometry.

Let me preface my remarks on this important branch of our subject by stating that at sea the datum line is always a horizontal line; spherical trigonometry is never used, and not one out of one thousand shipmasters understands spherical trigonometry.

In "Modern Science and Modern Thought," by Laing, we are informed, on page 54, that:

"These calculations...are as certain as those the nautical almanac, *based on the law of gravity WHICH* ENABLE SHIPS TO FIND THEIR WAY ACROSS THE PATHLESS OCEAN."

I have used the Nautical Almanac somewhat, but this is the first intimation I ever had that the few things it contains which are useful to the navigator, viz Sun's Declination, Equation of Time, Semi-diameter, and such-like, are "based on the law of gravity." Nor did I ever suspect that the calculations of the nautical almanac "enable ships to find their way across the pathless ocean," Such utter misstatements may suit the unthinking man to bolster up his theory, but they declare to the complete ignorance of the critic regarding practical navigation. A knowledge of the facts compels me to jettison the cargo to lighten the ship of such absurd misrepresentations. Sun's declination is the sun's distance north or south of the equator. Semi-diameter of a heavenly body is half the diameter which has to be added to the reading if the lower limb be taken, and substracted if the upper limb be observed, so as to get the altitude of the centre of the object. Equation of time is the difference between the real sun and the sun which the astronomer supposes to rise and set every day alike, called the mean sun. Except in taking lunars, these are all the elements required from the nautical almanac to work out an observation. In lunars the moon's parallax and right ascension are used and are given in the nautical almanac. The first of these depends on the moon's position and the second is reckoned from the first point of Aries, one of the zodiacal signs and a point in the heavens. None of these elements have anything whatever to do with the shape of the earth, and certainly none are in any way connected with the bogus "law of gravity." To a practical man, Mr. Laing's statement is both untrue and absurd.

Now let us go into the matter and see what actually is the case, and how and on what principle "ships find their way across the pathless ocean,"

I shall first deal with...

PLANE SAILING

In "A Primer of Navigation," by A. T. Flagg, M.A., page 65, we find the following:

"*Plain Sailing.* — When a ship, sails for el short distance on one course, *the earth is regarded as a plane or level surface*...The results obtained by this assumption, although not absolutely correct, are near enough in practice."

This does not look as if the "law of gravity" had a hand in the matter; neither, it must be confessed, does it appear that the Nautical Almanac has any connection with the subject. So while the reader is reflecting on what "figure" *a globe with a fine or level surface* would "cut," we may let the anchor for a brief space, so that A GI.OBE WITH A PLANE OR LEVEL SURFACE may be duly appreciated. If the reader cannot now find time to search Euclid and other works for the nondescript figure, he may find leisure some other time. But let us get the anchor aboard and proceed. In "Navigation and Nautical Astronomy," by J. R, Young, page 40, the author declares that:

"PLANE SAILING is usually defined to be the art of navigating a ship on the supposition that the earth is a plane. This definition is erroneous in the extreme, in all sailings the earth is regarded as what it really is, a sphere. Every case of sailing, from which the consideration of *longitude* is excluded, involves the principles of plane sailing; a name which merely implies that although the path of a ship is on a spherical surface, yet we may represent the length of this path by *a straight line on a plane surface*...Even when longitude enters into consideration, it is still with *the plane triangle only that we have to deal*...but as the investigation here given in the text shows, *the rules for plane sailing* WOULD EQUALLY HOLD GOOD THOUGH THE SURFACE WERE A PLANE."

It must be evident to everyone who understands what a triangle is, that the base of any such figure on a globe would be an arc of a circle, of which the centre would be the centre of the globe. Thus, instead of a PLANE triangle, the figure would contain one plane angle and two spherical angles. Hence, if the PLANE TRIANGLE is what we have to deal with, and such is the case, the base of I he triangle would be a straight line — the ocean. That all triangulation used at sea is *plane,* proves that the sea is a plane. The foregoing quotation states that a plane triangle is used for a spherical surface, but *"the rules for plane sailing would equally hold good though the surface were a plane."*

What fine reasoning. It is like saying that the rules for describing a circle are those used for drawing a square, but they would equally hold good though the figure were a square.

From Mr. Young, the mathematician, we ascend to Professor Evers, Doctor of Laws, surely he will be able to enlighten us. In his "Navigation in Theory and Practice," page 56, he tells us that:

"PLANE SAILING is sailing a ship, or making the arithmetical calculations for so doing, *on the assumption that* THE EARTH IS PERFECTLY FLAT...It is not a strictly correct supposition to take any part whatever of the earth's surface as a plane; yet when the vessel goes on *short voyages,* the results obtained by plane sailing will be *sufficiently correct to serve every useful purpose...*Plane sailing cannot always be advantageously employed, ALTHOUGH IN PRACTICE SCARCELY ANY OTHER RULES ARE USED BUT THOSE DERIVED FROM PLANE SAILING...*The great and serious objection to Plane Sailing is that longitude cannot be found by it* ACCURATELY, ALTHOUGH IN PRACTICE IT IS MORE FREQUENTLY FOUND BY IT THAN BY ANY OTHER METHOD."

This, I notice, extends the *principle* from "a short distance" by Flagg, to "short VOYAGES" by Evers. A voyage, then, may be completed by plane sailing. That is, the rules used in navigating the ship on a short voyage will he those that would "hold good though the surface were a plane," Flat surface all the way, that is it. But we are again confronted by "a globe with a plane or level surface;" clearly an impossibility. Now let us enquire how *long* the *short voyage* may be, to have "a plane surface all the way." In December, 1897, I met Captain Slocum on board the "Spray" This navigator told me that *he had sailed his little craft 33,000 miles by plane sailing.* Rather a LONG voyage, it must be admitted. A PLANE or LEVEL SURFACE for 33,000 miles, and yet the world a globe? To the prehistoric "man of science" at the North Pole, and the Darwinian Ape at the South Pole (?) of the astronomers' imaginary globe, with such a delusion.

Let it be put on lasting record that "in practice scarcely any other rules are used but those derived from plane sailing;" and that although "the great and serious objection to plane sailing ii that longitude cannot be found by it accurately," yet "IN PRACTICE IT IS MORE FREQUENTLY FOUND BY IT THAN BY ANY OTHER METHOD."

The only logical conclusion we can arrive at from the principles of Plane Sailing, *as furnished by its mathematical exponents,* is that IT PROVES THE WORLD A PLANE, and we know from actual practice that such is really the case.

But before saying adieu to this navigation proof, we must quote still further.

"Bergen's Navigation and Nautical Astronomy," 1st app., page 4, states:

"If the course and distance which a ship has sailed on the globe be given, the difference of latitude and departure may be found by the resolution of a right-angled plane triangle."

We have before seen that "a right-angled-plane-triangle" on a globular surface is impossible. So there is no need to comment on Captain Bergen's statement.

In "Navigation," by D. Wilson Barker, R.N.R., F.R.S.E., and W. Allingham, Plane Sailing is dealt with on page 29 as follows:

"We may now *assume* as an axiom that the shape of the earth somewhat resembles that of an orange. At one time people thought differently, *but no sane person to-day would venture to assert that our planet is merely an extended plane.* Still we shall not be far out IF WE IMAGINE that the small portion of the earth's surface with which we are concerned in Plane Sailing is ACTUALLY A PLANE. Hence, in Plane Sailing, regarding the small portion of the ocean with which we have to deal AS A FLAT SURFACE LIKE A SHEET OF PAPER, we have always A RIGHT ANGLE PLANE TRIANGLE TO WORK WITH."

These learned gentlemen say that no *sane* person to-day I would venture to assert that OUR PLANET is merely an extended plane; and yet they ask the reader to admit their sanity when they furnish data which prove the world to be a plane! Wonderful learning and profound philosophy that fit a plane triangle on to a spherical surface. Surely A GLOBE with a FLAT SURFACE LIKE A SHEET OF PAPER is a new figure, not found in Euclid or any of the works that deal with triangulation. We may well challenge the advocates of the globular theory to produce their globe with its plane or level surface like a sheet of paper, and be certain of their failure.

The spectre called "our planet" only requires to be planed (just a little *levelling*) to reduce its surface to a plane; and before we have finished the process the plane win be very plain indeed.

In the "Natal Mercury" of 14th March, 1898, the following example of 2,000 miles of plane sailing is furnished:

"Captain Moloney, of the "Briton," gave a representative of this journal particulars respecting the passage of the vessel through a dust storm on the way out. He said that they fell into the storm about 80 miles south of Madeira, and were in it for a distance of between 1,800 and 1,900 miles. They were without observations for 2,000 miles...so that they had to go over 2,000 miles on DEAD RECKONING."

This terrible sand-storm visited another ship, and planed off the supposed convexity of the water, so that plane sailing could be carried out and even *longitude* found by it for a further distance of 900 miles, as witness the "Natal Mercury" of 25th February, 1898.

"The experience met with by the 'Roslin Castle' on his homeward journey was most extraordinary. A sand-storm of unprecedented density enveloped the vessel, and rendered observation impossible for 900 miles. Madeira was reached by means of DEAD RECKONING."

Plane Sailing proves that the surface of water is a plane I or horizontal surface "like a sheet of paper," and in practice it is shown that this plane extends for many thousands of miles. Whether the voyage is outwards, as in the case of the "Briton"; or homewards, as in the case of the "Roslin Castle," makes no difference; thus showing that a "short voyage" to the Cape and back to England can be accomplished by plane sailing, flat water "like a sheet of paper" all the way.

The fact that water is flat like a sheet of paper (when undisturbed by wind and tide) is my "working anchor," and the powerful "ground tackle" of all those who reject the delusions of modern theoretical astronomy.

Prove water to be convex, and we will at once and forever recant and grant you anything you like to demand.

I will not waste time by quoting Mercator's, Middle Latitude, and Parallel Sailings, for they are merely plane sailing extended. Let us get on to what unthinking navigators believe to be a proof of the globularity of the world,

GREAT CIRCLE SAILING

Bergen's "Navigation," 1st appendix, page 9, states:

"Great circle sailing is founded on the principle that the shortest distance reckoned on the earth's surface between any two points, is the arc of the great circle intercepted between them."

The "arc of a circle" has undergone considerable planing when it leaves Mr. Wilson Barker's hands, for he informs us on page 95 that:

"We may ASSUME *as an axiom that the shortest distance between any two points is a* STRAIGHT LINE."

What, a *straight* line on a globular surface? Never, it is impossible. When it can be obtained, we surrender.

In "Navigation," by Rev. W. T. Read, M.A., page 51, the resource that is had to approximate great circle sailing is stated to be that:

"The vessel may be said to sail UPON THE SIDES of a *many-sided* PLANE FIGURE."

So, after all, the earth is not a globe, but a flat-surfaced many-sided plane figure—A POLYGON!

But how long is Mr. Wilson-Barker's STRAIGHT line? When the corner of the Polygon was reached *another straight line* would have to be followed, and another on the next side, and so on. Truly, these paste-board navigators are all "at sea" and don't know whether the ship is in the water or the water in the ship.

It is somewhat remarkable that J. R. Young, who so earnestly endeavours to support the globular hypothesis in his "plane sailing," does not even mention "great circle sailing" in his work already referred to. Plane sailing is sailing on a plane and there is not the remotest chance of proving convexity from it. If there be any semblance of globularity it can only be found in what is known as great circle sailing. There is, in reality, no such thing as sailing on a great circle, or on any circle except a flat one. On a globe, all circles that do not pass through the centre are called small circles, and to sail on one of them, it is said, is on the Rhomb-line or Mercator track, and the longest distance. But on any great circle — any circle that passes through the centre of the globe — the distance is said to be the shortest. The arc of the great circle between any two places on it is the shortest distance and is the great circle track.

I have already shown that water is level, "like a sheet of paper," as one author puts it. It is, therefore, quite impossible to sail a vessel on the globular arc of a circle, which is said to be done in following a great circle track. But Bergen's "Navigation" will help us. Page 247 of this work states that the great circle track may be found on a great circle chart by laying a *straight edge* on the ship's position and that of her destination, "the edge shows the track."

We simply ask for the globe that will bear the application of the straight edge. If it be argued that the great circle chart is merely a device for reducing the globular surface of the earth to a plane surface for the sake of simplicity, and that a curved surface can be represented by a straight line, we say it is impossible to represent a curved surface by a straight line and absurd to make the attempt, and we have already shown that water is flat, "like a sheet of paper"; we are, therefore, fully entitled to conclude that Captain Bergen's *straight edge* is applicable to a *straight surface* only. That this is what is really the case will appear later.

Rhomb-line sailing, which was mostly practised under certain conditions before Great Circle sailing was "discovered," is sailing the longest way round. The difference between the methods will be seen in the following; — Describe a circle, and mark any two places on it, say A and B. Let the circle be 12 miles in circumference, and A and B 3 miles apart. It is evident that if the rhomb-line from A to B be followed, the distance sailed will be 3 miles; but draw a straight line from A to B, and it will at once be seen that by following this track the distance will be shortened to 2¾ miles. *This straight line is the great circle track between A and B.* Or, if a piece of thread be drawn across a globe between any two places, the track thus obtained will be part of a great circle, and if this be transferred to a Great Circle chart IT WILL BE A STRAIGHT LINE, Therefore I conclude that great circle sailing is no discovery, for, had those who "discovered" it only perceived that the earth is a plane, they would have known that, on a plane surface, the shortest way is a straight line between two places.

Rhumb-line sailing between any two places on the same, parallel of latitude, would be sailing the ship east or west (as the case might be), *thus making a circular path*; whereas the great circle track would either be to the north or south of east or west, so as to get a straight line between the two places, which would be the shortest distance. It is surprising that anyone has claimed this as a discovery, and still more surprising to find anyone with a knowledge of navigation writing it down as proof of the earth's rotundity, THE GREAT CIRCLE TRACK ON A GLOBE ANSWERS TO A STRAIGHT LINE ON A PLANE SURFACE.' THE EARTH'S SURFACE IS A PLANE SURFACE, THEREFORE IT IS NO DISCOVERY TO FIND THE SHORTEST CUT TO BE THE MOST DIRECT ROUTE, ON THAT SURFACE.

Thus, great circle sailing, which is in reality rectilinear sailing, shows that the *chord* of the arc is a shorter distance than the arc, inasmuch as a *straight* line is shorter than a roundabout one can be. Let it be noted, however, that great circle courses are seldom followed on account of land and other impediments being in the way. Now we return to "Evers' Navigation." On page 192 we get his idea of great circle sailing as follows:

"The solution of problems in great circle sailing depending upon spherical trigonometry; hence to rightly comprehend whole subject, the student must be well versed in the solution of right angled and oblique spherical triangles."

When a Professor of Navigation says that spherical trigonometry is necessary to the practice of great circle sailing, of course the general reader believes the statement. Bui there is no truth in the statement all the same. I have already stated that spherical trigonometry is never used at sea, and that few navigators understand the subject. But there are few navigators who hold Board of Trade certificates that could not calculate the first and other great circle courses, the position of the vertex and the last course on a great circle track in a few minutes. How then can it be done by spherical trigonometry, if the calculators do not understand it? The answer is that it is done in every case by plane trigonometry. If the reader will procure a work on spherical trigonometry and one on *plane* trigonometry, he will see that the sines, co-sines, tangents, secants, &c., in relation to *the chord of an arc on a flat surface, are precisely the same as these quantities when taken in relation to the arc of a globular circle.* In Evers' "Navigation," pages 227 and 228, the "limitations of great circle sailing" are dealt with as follows:

"The difficulty in making (he calculations for great circle sailing are sufficient to deter the majority of practical men from adopting it. Again, as before intimated, many impediments, as islands, land, too high a latitude, &c., lies in its way. Several modifications to further extend its use, and mechanical methods already referred to, have been introduced. Theory and practice in this case are often widely separated. The sailing master has to take advantage of winds and currents, and considers how he shall make *the quickest passage,* which is not always the *shortest.* The best way to find out where the quickest passage can be made, is to lay down the great circle on a Mercator's chart, which has the winds and currents marked on it; then with the straight line on the chart joining the two places, first compare the two paths, *i.e.,* the Mercator's and great circle tracks, taking note of what currents of wind or water will assist the vessel; whichever offers the quickest passage is the best route, if not the shortest. Again, if by modifying the great circle track, by keeping to a lower latitude, the ship can be brought into currents in favour of the vessel, that will lie the best track. Although the greatest advantages of great circle sailing over the rhumb are obtained when sailing in high latitudes, yet, in consequence of the danger arising from ice and icebergs floating from the North Pole into the North Atlantic, and from the South Pole into the South Pacific and South Asiatic, navigators are unable to secure these advantages."

From page 193, Vol. I., of "Naval Science," we extract the following:

"In the passage from Panama to Australia, the rhumb track would entangle us in the Low Archipelago, in Dangerous Archipelago, and carry us into the very focus of coral reefs, atolls, lagoon islands, and sunken rocks, while the great circle route would take us clear of these dangers. On the other hand, the great circle track from Cape Horn to Cape of Good Hope (were there no other objections), would run the ship on one of the Sandwich group, while the rhumb course would carry her clear of such dangers."

In practice, therefore, it is clear that the advantages of what is known as great circle sailing, can seldom be secured, for the above reasons.

But if a vessel starts on a great circle course and sails on it one day, how is her position found by, plane triangulation only, and in every case, as I shall now proceed to show. The following example of "finding the latitude" from a meridian altitude of the sun is taken from Bergen's "Navigation," page 67:

EXAMPLE

I. 1865, March 4th, in longitude 4° 30' E., the observed meridian altitude of the sun's lower limb was 24° 49' 10", bearing south, index error —9' 50", height of eye 11 feet; required in the latitude.

	d. h. m. s.		° ′
Apparent time at ship, March ...	4 0 0 0	Longitude ...	4 30 E.
Longitude in time, East	− 0 18 0		4
Apparent time at Greenwich, March...	3 23 42 0		60) 18 0
		Long. in time	0 18 0
Hours and decimals of hours ...	23·7		

	° ′ ″		″
Sun's declination at noon, March 3rd	6 41 28 S.—	Diff. for one hour	57·55
Correction	− 22 44	Hours, &c.	23 ·7
			40,285
Sun's reduced declination ...	6 18 44 S.		17,265
			11,510
		6,0)	136,3·935
		Correction ...	22 44

	° ′ ″
Observed altitude, sun's lower limb	24 49 10 S.
Index error...	— 9 50
	24 39 20
Dip., Table V., for 11 feet	— 3 16
Apparent altitude, sun's lower limb	24 36 4
Sun's corr., Table VII.	— 1 57
	24 34 7
Sun's semi-diameter, page II., Naut. Almanac	+ 16 10 or Table VIII.
True altitude, sun's centre	24 50 17
	90 0 0
Sun's zenith distance	65 9 43 N.
Declination...	6 18 44 S.
Latitude	58 50 59 N.

The sextant, or quadrant, is an instrument used to measure the altitude of any object above the surface of the earth. The former will measure angles up to 120°. The latter instrument only measures up to 90° — hence a quadrant Except in taking a lunar, where two heavenly bodies are at a greater angular distance than 90°, the quadrant will do as well as the sextant.

Having previously adjusted the instrument, with the sextant bring down the image of the sun to the horizon at noon, and note the reading. In the example before us, the instrument had an error, which is allowed for. If the observer's eye were at water-level, there would be nothing to deduct for "height of eye" (erroneously styled "dip"). But as the eye is always above the water, and *consequently a greater angle* is obtained, an amount must be deducted to give *the reading that would have been obtained with the eye at water level,* that being the datum line. Therefore, "height of eye" must be deducted.

With the eye at water level at one angle and the sun at water level at the other, the line joining them is the base of the triangle — a straight line, of which we have already heard so much. But if water be convex, when the height of eye is deducted and the observation reduced to the datum line—the sea, then the eye and the sun are both at the surface of the convex water, consequently the base of the triangle is the arc of the circle between the two points, and another allowance must be made to reduce this arc of a circle to a straight line, in order to determine the true angle of the plane triangle. That this is not only never done, but that no work on Navigation ever published makes the slightest reference to the need for such a correction, and that all triangulation in Navigation is *plane,* proves incontestably that the surface of the ocean is a plane surface.

Having deducted height of eye, deduct the refraction (which raises the image of an object above its true position) if any exists, and the result is the true altitude. Then, if the lower limb of the sun be observed, add half the diameter so as to get the true altitude of sun's centre. Then a further fact requires to be noticed. The sun, when on the equator, that is, when it has no declination, makes a right angle with the ocean and land at all points on the equator. This fact and horizontal water are the main data in observations for finding the ship's position at sea. Deduct what has now been arrived at from the right angle (90°), the remainder is the sun's zenith distance. Then, if the sun had no declination, the zenith distance would be the latitude; but as the sun in the present case is *south* of the equator and the ship in *north* latitude, the declination (sun's distance from the equator) has to be substracted to give the latitude. The declination, I may notice, is the reduced declination. That is, the declination reduced to the longitude of the ship. As the sun only makes a perfectly circular path about four times in a year, his path being eccentric at all other times; it is required to know the variation of the declination, the eccentric above referred to being a spiral or corkscrew movement. If at Greenwich the declination is a given amount, and the variation for one hour be known, we only require to know how many hours the ship is east or west of Greenwich to know by how much to multiply the variation, to get the amount to be added if declination be increasing, or subtracted if it be decreasing.

Much time could be saved by the use of an instrument pivoted vertically and supported by four legs with gimballs and weighted with lead to preserve the instrument vertical; with a sight to take the angle of the sun, that is, its difference from the vertical (90°), which, with the declination applied, would give the latitude in a few minutes. In all these quantities there is not the remotest reference to the rotundity of the earth, but the very opposite, as the datum line—flat water, is one of the main factors.

In finding the longitude also, the same method of triangulation is used. If the surface of the ocean be globular, *there are no rules laid down for calculating on that basis.*

The allowance for convexity is never made, and it would be impossible to allow for it, as in clear weather the horizon is distant, while in thick weather it is very near. To reduce the curved base of a spherical triangle to a straight line of a plane triangle is an impossibility, because the factors are unknown and in the nature of the case, never can be known.

The whole of navigation, therefore, furnishes strong evidence that the world is not the globe of astronomical speculation and popular credulity, but a plane figure.

The base of the triangle is always the straight line projected from the observer; and a straight line requires a flat or horizontal surface for its projection.

It is commonly supposed that meridians of longitude south of the equator, converge to a common centre, as they do in north latitudes. If this were so, the allowances to be made for the longitudes being shorter as the south was approached would show the ship to be in her true position.

Captain Woodside, of the American barkentine *Echo,* at Capetown, in June, 1898, says that on 12th January, 1896, being without observation for two days and sailing a straight course at 250 miles a day, he expected to be about 100 miles to the southward, and a long way to the eastward of Gough Island, in latitude 40° south; but was startled to find the ship making straight for the island, and barely escaped shipwreck. This proves that although the usual allowance for shorter longitudes in the south had been made, the ship's position was not known. There must, therefore, be something wrong with the assumed length of degrees of longitude in the south. In the case above referred to, the ship was going to the eastward, and had an allowance in excess of the usual length of a degree of longitude been made, so as to correspond to what the length of degrees are at 40° south latitude, the ship's longitude would have been known. That it was not known proves that degrees are longer at 40° south latitude than at the same latitude north of the equator.

In "South Sea Voyages," by Sir James C. Ross, page 37, it is stated:

> "By our observations at noon we found ourselves 58 miles to the eastward of our reckoning in two days."

And in a "Voyage towards the South Pole," by Captain James Weddell, we find the following:

"At noon in latitude 65° 53' South our chronometers gave 44 miles more *westing* than the log in three days."

Lieutenant Wilkes informs us that:

"In less than 18 hours he was 20 miles to the *east* of his reckoning in latitude 54° 20' South."

The discrepancies in the above cases were attributed to currents, *whether the course of the ship was westerly or easterly,* which could not possibly be the case. These navigators, believing the world to be globular could not imagine any other way of accounting for the discrepancies between longitude by "dead reckoning," making allowance for the supposed shorter longitudes, and that obtained by observation. The explanation is that the world diverges as the south is approached, instead of converging, as the theory teaches.

It has also been shown under "Distances" chapter of this work, that at latitude 32° south, the distance round the world is about 23,000 statute miles; at latitude 35½° south, the distance round is over 25,000 miles; and still further south, at latitude 37½° south, the distance is 25,500 miles, about. These distances, obtained from ship's logs, cannot disputed; and are altogether against the theory of the earth's rotundity. By purely practical data, apart from any theory, it is shown that the world diverges to the south, and that, therefore, it cannot be a globe.

Twenty-Five – The Pendulum

Sir Robert Ball, in his "Story of the Heavens," page 177, says:

"We find that by observing the swing of a Pendulum at different parts of the earth, we are enabled to determine the shape of our globe."

This is perhaps one of the greatest fallacies of the globular school, and when looked at without prejudice, is sheer nonsense. A Vibrating Pendulum on a globe with various movements would move with the globe, and could not by any possibility record the movement of the globe to which its supports were fastened.

The following is from "Noad's Lectures on Chemistry," page 4:

"All the solid bodies with which we are surrounded constantly undergoing changes of bulk, corresponding to the variations of temperature...The expansions and contraction of metals by heat and cold form subjects of serious and careful attention to chronometer makers, as will appear by the following statements:—The length of the pendulum vibrating seconds, *in vacuo*, in the latitude of London (50° 31' 8" north) at the level of the sea, and at the temperature of 62° Fahr. has been ascertained with the greatest precision to be 39.13929 inches.

Now, as the metal of which it is composed is *constantly* subject to variations of temperature it cannot but happen that its *length* is constantly varying, and when it is further stated that if the 'bob' be let down 1.100 of an inch, the clock will lose ten seconds in twenty-four hours; that the elongation of 1,000 of an inch will cause it to lose one second per day; and that change of temperature equal to 30° Fahr. will alter its length 1.5000 part, and occasion an error in the rate of going of eight seconds per day, it will appear evident that some plan must be devised for obviating so serious an inconvenience."

In the "Figure of the Earth," by J. Von Gumpach, we are informed as follows:

"General Sabine himself," relates Captain Foster, "was furnished with two invariable pendulums of precisely the same form and construction as those which had been employed by Captain Kater and myself. Both pendulums were vibrated at all the stations, but FROM SOME CAUSE, which Mr. Bailey was UNABLE TO EXPLAIN, the observations with one of them were SO DISCORDANT at South Shetland as to REQUIRE THEIR REJECTION."

The *English Mechanic* of 23rd October, 1896, has the following, signed by a fellow of the Royal Astronomical Society:

"In reply to 'Foucault's Pendulum' (Query 89,090, p. 192), the plane of oscillation of the pendulum in latitude 5° would rotate in a retrograde direction at the rate of only 1.307336° per hour; in other words, it would take 11.4737 days, or about 11½ days, to complete its rotation. Hence, while it might theoretically be employed to show the earth's rotation, IN PRACTICE IT MUST FAIL TO DO SO."

"Iconoclast," writing in the *Earth Review,* for April-June, 1897, says (*inter alia*):

"The so-called pendulum proof of the world's assumed rotation was obliged to be renounced years ago as worthless, by who were in the best possible position to judge, as these few of numerous extracts show: 'The first position of these theorists is, that in a complete vacuum, beyond the sphere of the earth's atmosphere, a pendulum will continue to oscillate in one and the same original plane- On that *supposition* their *whole theory is founded.* In making this supposition the fact is overlooked that there is *no vibratory motion* unless through atmospheric resistance, or by force opposing impulse. Perpetual in rectilinear motion may be imagined, as in the corpuscular theory of light; circular motion may be also found in the planetary systems; and parabolic and hyperbolic motion ta those of comets; but vibration is artificial and of limited duration. No body in nature returns the same road it went, unless artificially constrained to do so. The supposition of a permanent vibratory motion, such as is presumed in the theory advanced is *unfounded in fact* and absurd in idea; and the whole affair of this proclaimed discovery falls to the ground.'

"Liverpool Mercury," May 33rd. "T."

Again, in the same month, appears the following;

"A scientific gentleman in Dundee recently tried the pendulum experiment, and concludes by saying, 'That the pendulum is capable of showing the earth's motion, I regard as a gross delusion...'"

Again, another asserts. "He and others had made many pendulum experiments, and had discovered that the plane of vibration had nothing whatever to do with the meridian longitude, nor with the earth's motion..."

In many instances experiments have however not even shown a change in the plane of oscillation of the pendulum; in others the alteration has been in the wrong direction; in fact, in numerous instances, the rates and directions have been altogether opposite to that which the theory indicated; a notable illustration of this was given publicly by the Rev. H. M. Jones, F.R.A.S., in 1851, at the Library Hall of the Manchester Athenaeum, where the diurnal rotation of the earth was to be attempted to be demonstrated by delicately adjusted Pendulum; after giving, at length, a minute description of the arrangements and apparatus, we come to the admission, that the pendulum, on being released, travelled over a measured space in seven minutes, whereas, according to the theory, it ought to have taken fifteen minutes, or more, to accomplish the distance; and remember, this great difference was made without any allowance being made for the resistance of the air, which would be considerable. Anyone can verify this account by referring to the "Manchester Examiner Supplement" of May 34th, 1851.

By referring to "The Figure of the Earth," by J. Von Gumpach, 2nd edit., 1862, on pp. 229 to 244, results will be seen of Sixty-seven experiments with the Pendulum, made in every latitude North and Twenty-nine South of the Equator, by Captains Foster and Kayter, and General Sabine, all of which are admitted to be absolutely worthless for proving anything regarding the assumed motion of The World through space.

If such testimony is not enough to make Pendulum-proof worshippers think they must either be as bigoted as it is possible to conceive, or as thick in the cranium as their globe."

The vibrations of a pendulum, therefore, whatever value they may have in determining something as yet unknown, can have nothing to do with supposed motions of the earth, and must be relinquished by every thinking man.

Twenty-Six - Plurality of Worlds

Sir David Brewster, in his "More Worlds than One," says:

"It was not until the form and size and motions of the earth were known and till the condition of the other planets was found to be the same, that *analogy compelled us to believe that* THESE PLANETS MUST HE INHABITED LIKE OUR OWN..." "The doctrine was maintained by almost all the distinguished astronomers and writers who have flourished SINCE THE TRUE FIGURE OF THE EARTH WAS DETERMINED..." "Under these circumstances the scientific world has been greatly surprised at the appearance of a work entitled 'Of a Plurality of Worlds', the object of which, like that of Maxwell, is to prove that our earth is the only inhabited world in the universe, while its direct tendency is to ridicule and bring into contempt the grand discoveries in sidereal astronomy by which the list century has been distinguished."

In "Sun, Moon, and Stars," by A. Gibeme, page 10, the following is found:

"Just as our sun is a star, and stars are suns, so our earth or world is a planet, and planets are worlds." "The planets are worlds, more or less like the world we live in."

And in his "History of the Conflict between Religion and Science," Dr. Draper tells us that:

"If each of the countless myriads of stars was a sun surrounded by revolving globes, peopled with responsible beings like ourselves; if we had fallen so easily and had been redeemed at so stupendous a price as the death of the Son of God, how was it with them? Of them were there none who had fallen or might fall like us? Where, then, for them, could a Saviour be found?"

IF the world be the globe of popular belief; IF the sun be a million and a half times the size of the earth-globe and about 100,000,000 miles distant from it; IF the stars are worlds and suns, distant many millions of miles and vastly larger than even our own sun; IF the earth was a piece of molten rock shot off from the sun; IF the moon was a piece fractured off from the earth; THEN it is a very proper question to ask, "Are these mighty globes in space Inhabited?" If so, are their inhabitants of a higher or lower order than the inhabitants of this globe?

Sir D. Brewster says that the plurality of worlds rests upon a few simple facts, and the foregoing are said to be dome of these facts; but it was not till the form and size and motions of the earth were known ANALOGY compelled the belief that the planets must be *inhabited worlds like ours.* I have already shown that those who believe modern astronomy, and, by consequence, the plurality of worlds, are of all men most ignorant as to the shape of the world they live on; that it has none of the terrific notions attributed to | it; and that, unlike celestial bodies, it is a terrestial structure, a stationary plane.

The following quotation from "A Treatise on Astronomy," by E. Henderson, L.L.D., F.R.A.S., shows that the whole of this supposed analogy is based upon conjectures, and must therefore be rejected.

"The great *probability* is that every star is a SUN far surpassing ours in magnitude and splendour; they all shine by their own native light...What a most powerful SUN that little star Vega must be, when it is 53,977 times larger than our sun...The stars thus being SUPPOSED to be suns it is EXTREMELY PROBABLE that they are the centres of *other systems of worlds,* round which may revolve a numerous retinue of planets and satellites. *Therefore, there must be plurality of suns,* A PLURALITY OF WORLDS."

The plurality of worlds, therefore, is based on assumptions so contrary to known possibilities, that the "grand idea" must be thrown into the waste-paper basket.

The supposed great distance of the sun from the earth is the main cause of the delusions of the learned as to the called worlds above us being inhabited.

This distance is based on a fictitious idea, that of the revolution of the earth round the sun; which I have already shown to be unconditionally false. The sun is a small body of light and near the earth, therefore all the star distances are wrong, their sizes and all other suppositions.

The plurality of worlds is only the logical sequence of belief in the earth being a rapidly revolving globe. But this has been shown to be ridiculous in the extreme. Evidence, apart from any theory has been presented which entirely nullifies such an assumption, and renders it absurd; showing that such an unnatural idea has not a vestige of natural fact to support it. The grand doctrine of the plurality of worlds, therefore, like all the other grand doctrines of modern astronomy, must be consigned to oblivion. When it can be shown that this world is a globe and by what known principle the inhabitants can hang on to the swinging ball, like the house fly crawls along the ceiling, it will be quite time enough to talk about the plurality of worlds.

Twenty-Seven - The Planets

If all that astronomers have to say about themselves were correct, they would be about the wisest as well as the cleverest men that ever existed. There are not many modest men among them, but the quotation which follows is about the most *immodest* that can well be found. It is taken from "The Story of the Heavens," from which we have quoted so frequently:

"Astronomers have *taken an inventory of each of the Planets. They have measured their distances, the shapes of their orbits, and the positions of these orbits,* their times of revolution, and in the cases of all the large planets their sizes and THEIR WEIGHTS." ... "*It is not an easy matter to weigh the earth* on which we stand. How, then, can we weigh a mighty planet vastly larger than the earth, and distant from us by some hundreds of millions of miles. *Truly this is a bold problem.* Yet the intellectual resources of man have proved sufficient to achieve this feat of celestial engineering...ALL SUCH INVESTIGATIONS ARE BASED ON UNIVERSAL GRAVITATION." "A foot-rule placed at a distance of 40 miles subtends an angle of a second, and it is surely a delicate achievement *to measure the place of a planet and feel confident that no error greater than this can have intruded into our result."*

The uninitiated reader may gape with wonder when reading these and such-like absurdities, but we shall see presently how great the errors are, which have intruded into the calculations of the wise men. But first, as to the basis of the whole of these supposed achievements of scientific celestial engineers. IT IS SAID TO BE FOUNDED ON UNIVERSAL GRAVITATION, which we have proved to be, like most other statements of the wise men, A MYTH.

Now as to the *small errors.* "Our Place among Infinities," by R. A. Proctor, page 166, informs us that:

"If the error in the estimate of the sun's distance appears startling, what will be thought of an error which must be estimated by millions of millions of miles? If the estimate of the star's distance were correct, the distance of Sirius would amount to about 130 millions of millions of miles; THE CORRECTED ESTIMATE is as above mentioned 80 millions of millions of miles."

Thus, gentle reader, the very exact men of science acknowledge an error of 80,000,000,000,000 miles. What do you think of this? If a man thinks at all, he must think that these wise men know nothing at all about the stars and distances of the heavenly bodies, which must by necessity be lighter than air or they would fall to the earth, as anything that is heavier than air does. When such errors are unblushingly admitted and the figures based on the law of gravitation, the results arrived at must be as mythical as we have seen the law of gravitation to be.

T. G. Ferguson, in the *Earth Review* for September, 1894, says:

"Let us now glance at their theories about the Planets...Saturn's mean distance from the sun. as given in the 'Story of the Heavens,' is 884,000,000 miles, and the diameter 71,000 miles, Professor Lockyer gives its distance as 890,000,000 miles; a difference of 4,000,000 miles. Professor Olmstead gives Saturn's distance from the sun as 890,000,000 miles, and the diameter 79.000 miles. Others could be quoted equally at variance. WHERE, WE ASK, IS THE ACCURACY OF THIS 'MOST EXACT OF SCIENCES.'"

Were it necessary we could fill a good many pages with the errors of this exact science; enough has been said to prove to the thinking man that the wise men we have quoted know no more about the planets, their sizes, weights, and distances than did Hodge when, after having listened to a very learned discourse about the starry heavens, he was asked what he thought of the marvellous fact that light had taken from creation to travel from some of the fixed stars to the earth, he exclaimed, *"Law, Sir, what a big lie it do be, to be sure."*

Twenty-Eight - On Parallel Lines

The term "parallel" signifies equidistant, hence the self-evident truth that "parallel lines never meet." Because they are at equal distance from each other, they can never meet, no matter how far they may be prolonged. If lines do meet when prolonged, it is because they are not parallel or equidistant from each other. The above is so well-known that it seems at first sight a waste of words to re-state it, but the following quotations will .show the necessity of emphasing even self-evident truths.

"Some Unrecognised Laws of Nature," by I. Singer and L. H. Berens, page 11, contains the following:

"We suspend two plumb lines at a convenient distance and then measure their distances from each other at both ends. The most delicate measurement at present possible would demonstrate — as far as this is possible by direct observation — that the two lines are parallel to each other. By the aid of the abstract axiom that parallel lines if extended indefinitely would never meet, we would draw the inevitable inference that two such plumb lines, if indefinitely extended would never meet. This conclusion would seem obvious and inevitable; yet the student of to-day knows it to be false. But his knowledge is not due to direct observation, but to his acquaintance with the fact that the earth is round, and that plumb lines at any part of the earth are at right angles to the horizon."

I have not read one work on Astronomy which does not require on enormous amount of credulity if the reader is to accept as truth whatever is presented to him, but the above quotation will equal anything anywhere for the amount of credulity it pre-supposes the reader to be possessed of. By direct observation and experiment it is proved that parallel lines can never meet, being equidistant from each other. Yet the student after having proved the truth of the proposition, knows it to be false!!! Parallel lines can never meet, because they are parallel, no matter what the figure of the world may be. The same work, on page 13, states:

"To the man who conceived the earth as a flat expanse nothing could be more conclusive than that plumb lines were strictly parallel...But notwithstanding such direct and positive evidence, the student of to-day disbelieves this conclusion, and that not because he has any direct evidence to the contrary, but because it conflicts with the now established fact that our earth is a sphere. His evidence is not due to direct observation, but is circumstantial depending on a concatenation of inferences."

It would be difficult to conceive anything more opposed to reason and common-sense than the foregoing. One fact is done to death by what is said to be another fact, which is manifestly impossible, and one marvels how educated men can lend themselves to support what their own experience condemns. The same work, continuing on page 15, says:

"The reason why 'parallel lines never meet' is because we conceive them so and because as soon as lines approach towards each other we no longer call them parallel."

"This conclusion will enable us to understand why of two such conclusions—as: (1) plumb lines are parallel; (2) plumb lines are convergent,—we accept the latter, though based on a long chain of inferences, as against the former which is the result of actual observation."

Now, the most amateur draughtsman knows that parallel lines are not parallel, "because we conceive them so," but because they are equidistant from each other, and, therefore, can never meet if extended indefinitely. So that the gifted authors of the work from which I quote have actually to mentally destroy a fact and to deny self-evident truth in order to support what depends on a "concatenation of inferences." The "long chain of inferences" has to be accepted as truth as against the result of actual observation! If plumb lines are parallel, how can they be convergent? Truly, this globe theory depends for its support on the stultification of common-sense, the free run of the imagination and the dethronement of the reasoning powers. According to the globular hypothesis, parallel perpendiculars are impossible, yet any builder will admit that a house is a mass of parallel perpendiculars.

"Mensuration," by T. Baker, C.E., page 1, gives the definition of parallel lines as:

"Parallel lines are always at the same distance, and never meet when prolonged."

"The authors of "Some Unrecognised Laws of Nature" "have gone to strange lengths to support the fiction of a globe world. It never occurred to them that their experiment proving plumb lines to be parallel proved also that the world is not a sphere but a plane!

Twenty-Nine - Railways

In projecting railways on a globe, the datum line would be the arc of a circle corresponding to the latitude of the place. That the datum line for railway projections is always a horizontal line, proves that the general configuration of the world is horizontal. To support the globe theory, the gentlemen of the observatories should call upon the surveyor to prove that he allows the necessary amount for "curvature." But this is what the learned men dare not do, as it is well-known that the allowance for the supposed curvature is never made. 'In the session of the British Parliament for 1862, Order No. 44 states;

"That the section be drawn to fie same HORIZONTAL scale as the plan, and to a vertical scale of not less than to every one hundred feet, and shall show the surface of the ground marked on the plan, the intended level of the proposed work, the height of every embankment, and the depth of every cutting, and a DATUM HORIZONTAL LINE *which shall be the same throughout the whole length of the work...*"

In the *Birmingham Weekly Mercury*, of 15th February, 1890, "Surveyor" writes as follows:

"'An Engineer of thirty years standing' wrote to a Magazine in 1874 quoting the following sentence as the result of his experience in the construction of railways, more especially:— 'I am thoroughly acquainted with the theory and practice of civil engineering. However bigoted some of our professors may be in the theory of surveying according to the prescribed rules, yet it is well known amongst us that such theoretical measurement are *incapable of any practical illustration.* All our locomotives are designed to run on what may be regarded as TRUE LEVELS or FLATS. There are, of course, partial inclines or gradients here and there, but they are always accurately defined and *must be carefully traversed.* But anything approaching to eight inches in the mile, increasing as the square of the distance, COULD NOT BE WORKED BY ANY ENGINE THAT WAS EVER YET CONSTRUCTED. Taking one station with another all over England and Scotland, *it may be stated that all the platforms are* ON THE SAME RELATIVE LEVEL. The distance between the Eastern and Western coasts of England may be set down as 300 miles. If the prescribed curvature was indeed as represented, the central stations at Rugby or Warwick ought to be close upon three miles higher than a chord drawn from the two extremities. If such was the case there is not a driver or stoker within the Kingdom that would be found to take charge of the train...We can only laugh at those of your readers who seriously give us credit for such venturesome exploits, as running trains round spherical curves. Horizontal curves on levels are dangerous enough, vertical curves would be a thousand times worse, and with our rolling stock constructed as at present physically impossible. *There are several other reasons why such locomotion on iron rails would be* AS IMPRACTICABLE AS CARRYING THE TRAINS THROUGH THE AIR."

This important evidence by a practical man, may be supplemented by the following from W. Winckler, M.I.C.E., in the *Earth Review* for October, 1893:

"As an engineer of many years standing, I say that this absurd allowance is only permitted in school books. No engineer would dream of allowing anything of the kind. I have projected many miles of railways and many more of canals and tie allowance has not even been thought of, much less allowed for. This allowance for curvature means this — that it is 8" for the first mile of a canal, and increasing at the ratio by the square of the distance in miles; thus a small navigable canal for boats, say 30 miles long, will have, by the above rule an allowance for curvature of 600 feet. Think of that and *then please credit engineers as not being quite such fools.* Nothing of the sort is allowed. I must, I however, state that college astronomers have made the student engineer to think that in his method of levelling what is known as the "backsight" cancels any curvature by his "foresight", and so on. It is only a theory, and if our method of levelling cancels the obligation of making this allowance, we shan't quarrel with them — it does no damage to our projects when we get into practice, *but we no more think of allowing 600 feet for a line of 30 miles of railway or canal,* than of wasting our time trying to square the circle."

Astronomers know full well that it is no use appealing to the engineers, as their testimony is dead against the globular theory, although many of them believe in it all the same; but I never met one who said that he ever made the allowance said to bc necessary for projecting railways on the surface of "our tiny globe." In "Theoretical Astronomy," page 46, the author tells us that:

"Mr. J. C. Bourne, in his magnificent work called 'The History of the Great Western Railway' ...which is more than 118 miles long... '*the whole line* with the exception of the inclined planes, *may be regarded practically as level.*'"

One hundred and eighteen miles of LEVEL railway, and yet the surface on which it is projected a globe? Impossible. It cannot be.

Early in 1898 I met Mr. Hughes, chief officer of the steamer "City of Lincoln." This gentleman told me he had projected thousands of miles of level railway in South America, and never heard of any allowance for curvature being made. On one occasion he surveyed over one thousand miles of railway which was a perfect straight line all the way. It is well known that in the Argentine Republic and other parts of South America, there are railways thousands of miles long without curve or gradient. In the "Cruise of the Falcon," by that intrepid traveller and navigator, E. F. Knight, it is stated in Vol. 2, pages 1 and 2:

"From Tucuman to Cordova we were carried by the Government railway." "There are no curves on the way, the rails being carried in ONE PERFECTLY STRAIGHT LINE ACROSS THE LEVEL PLAINS."

In projecting railways, the world is acknowledged to be a plane, and if it were a globe the rules of projection have yet to be discovered. Level railways prove a level world, to the utter confusion of the globular school of impractical men with high salaries and little brains.

Thirty - Rivers

Rivers run DOWN to the sea because of the inclination of their beds. Rising at an altitude above sea-level, in some cases thousands of feet above the sea, they follow the easiest route to their level — the sea. The "Parana" and "Paraguay" in South America are navigable for over 2,000 miles, and their waters run the same way until they find their level of stability, where the sea tides begin. But if the world be a globe, the "Amazon" in South America that flows always in an easterly direction, would sometimes be running uphill and sometimes down, according to the movement of the globe. Then the "Congo" in West Africa, that always pursues a westerly course to the sea, it would in the same manner be running alternately up and down, When that point of the globe exactly between them was up, they would both be running up, although in opposite directions; and when the globe took half a turn, they would both be running down! We know from practical experiment that water will find its level, and cannot by any possibility remain other than level, or flat, or horizontal—whatever term may be used to express the idea. It is therefore quite out of the range of possibility that rivers could do as they would have to do on a globe.

Thirty-One - Ridicule

Sir D. Brewster speaks of a work, "the direct tendency of which was to ridicule and bring into contempt the grand discoveries in sidereal astronomy by which the last century has been distinguished."

No wonder that supposed discoveries, which are really only baseless assumptions, should call forth volumes to bring contempt and ridicule upon the impossible theories by which the last century speculators made themselves ridiculous.

The "Birmingham Daily Mail," of 25th November, 1893, states that:

"The astronomers arranged for a grand display of fireworks on Thursday night, the 23rd inst,, but the ungrateful fireworks did not appear. The showmen now take refuge in the clouds which shrouded the sky and say the fireworks were there only they could not be seen...It is believed that throughout the night we were careering through a storm of red-hot meteorites, the fragments of a comet smashed by a blundering planet some forty years ago..."

When newspapers ridicule the thing it must be very absurd, for they generally side with the professional men, the "Morning Leader," of 21st November, 1892, has the following:

A VERY DISTINGUISHED VISITOR

We have no desire to unduly alarm our readers, but our duty to the public compels us to announce *that to-night a collision may be expected between the earth and a comet.* The notice we give is somewhat short, so short indeed that if the worst comes to the worst, some distant readers may have barely learned the fact before the shock gives it an emphatic confirmation. The Rev. M. Baxter has somehow or other over-looked this noteworthy prediction, an oversight possibly accounted for by his feverish desire to discover some unfortunate individual who may be publicly described as "The Beast" without running foul of the law of libel...

Just at present it is perhaps risky to speak disrespectfully of comets, but it is undeniable that they are chiefly distinguished by their eccentricity. They resemble in no small degree political parties. They consist of a definite point or nucleus, with a remarkably nebulous tail preceding or following the nucleus. The tail precedes the nucleus when the comet has passed the perihelion and is receding from the sun, and it follows it when the sun is approached. That is to say, it is always to the front in a retreat and in the rear in an attack. As with the humble members of political parties, its distinguishing feature is prudence. Nor does the resemblance end here, for astronomers assure us that comets' tails ace, noted for their extreme tenuity. Stars which the slightest fog completely obscures shine through *millions (?)* of miles of their transparent material. In the same way it is easy to see through the motives and tactics of the political hanger-on. The nucleus is really the only part of a comet which need be noticed by practical men. The vaporous tails have frequently come within the earth's *attraction (?)* and have been absorbed into its atmosphere, just as the Literal Unionists have been "merged" into the Tory party. *Whether the effect of* the absorption of a comet's tail into our atmosphere has been salubrious or deleterious, or even if the event has had any perceptible influence at all, is only a matter of speculation among the learned. This extremely negative result resembles the action of homeopathic medicines upon the human frame — at least, as described by allopaths. The moral seems to be that the world will be wise if it carefully avoids the nucleus to-night and collides simply with the tall. "Run into something cheap," shouted the economical peer to his coachman when his horses bolted down Piccadilly.

Mankind has received comets in various moods. Sometimes they have been hailed with rapturous welcome. They have been supposed to herald a superior wine vintage. The produce of 1811 and of 1858 was specially announced as "comet wines," and topers declared that it was very good. On the other hand, these eccentric heavenly bodies have been regarded with hatred and terror. They were included in a very uncomplimentary prayer in the year 1456. The Turks had just captured Constantinople, and it was feared that they would soon overrun Europe. A comet was hovering about at the time, and the pious of the day added to the Ave Maria the following supplication: "Lord save us from the devil, the Turk, and the comet," It is strange that at the end of the nineteenth century we could be threatened by the same three influences. The first seems destined to be always with us, the second will haunt us until the Eastern Question is really settled, and the third threatens la mend or end us to-night.

"Reynold's Newspaper," of 27th November, 1892, has the following:

"A Dalziel Telegram, dated Philadelphia, November 24, says Professor Synder, Instructor of Astronomy in the High School here, states that the earth last night collided with a comet in tie Andromeda group and shattered it to pieces. This theory is *said* to receive confirmation by news from Illinois and other Slates, where there was a great fall of meteors. These are *supposed* to be the remains of the defunct comet."

The "Natal Mercury," of 30th August, 1898, says:

"To shift the axis of the earth from the poles to the equator M. Fouche, who has been working for years at the problem, says it is perfectly possible. It is only necessary to accumulate a sufficient quantity of material to one point of the equator, and the earth will 'turn turtle,' and continue its rotation at right angles to its present turning, while climatic, zoological, and social changes would ensue. The question is, how much material? M. Fouche answers 66 sextillions of tons. With all the resources of steam, the operation could not occupy less than two million years."

Thirty-One - The Sun

R. Russell, in his "Wonders of the Sun, Moon, and Stars," tells us, on page 86, that:

"The modern theory of the solar system maintains that the sun is comparatively motionless in the centre."

Our own senses testify against this delusion. No one ever yet felt or saw the earth careering through space at the terrific rates it is credited with, but everyone who is not blind can see the sun move. But the matter can be tested. It may be known for certain whether the sun moves or not. Take a school globe and place a stile on the semicircle that holds it in position. Cause the globe to rotate against a lamp on a table, and you will find that the shadow left on the globe is always parallel to the equator, at whatever angle you may incline the globe. Further, let the stile be of sufficient length to allow the shadow to fall on to a surface, moving the globe towards the lamp, and the shadow will be a straight line. If, therefore, the shadow left on the earth by the sun be a straight line, then undoubtedly the sun is stationary. Drive a stake into the ground in such a position as to expose it to the sun for the greater part of a day — the whole day if possible. Mark the end of the shadow every quarter of an hour, and you will find that the marks form part of an elongated curve, clearly proving that the sun moves over a stationary earth.

Thirty-Two - Sun's Distance

R. A. Proctor, in his work "The Sun," says that:

"The determination of the sun's distance is not only an important problem of general astronomy, but it may be regarded as THE VERY FOUNDATION OF ALL OUR RESEARCHES."

In R. Russell's "Story of the Solar System," we are informed that:

"The mean distance of the earth from the sun may be taken to be about 93 million miles, and this distance is employed by astronomers as the unit by which most other long celestial distances are reckoned."

Seeing then, that everything depends on the knowledge of the sun's distance from the earth, it is no wonder that it is regarded as one of the prime problems in astronomy. Surely this will be right; if not, all the rest will be wrong. Let us see what the wise men say. Let us see with what concurrence of "precise" calculations they agree as to this admittedly very important matter.

Sir R. Ball tells us that "the spirit of astronomical enquiry is NOT SATISFIED WITH APPROXIMATE RESULTS."

I have already quoted R. Russell as stating that the distance of the sun from the earth is 93 million miles.

In the "History of the Conflict between Religion and Science," by J. W. Draper, pages 173 and 174 inform us as follows on this important matter:

"In the time of Copernicus it was supposed that the sun's distance could not exceed live million miles, and indeed there were many who thought that estimate very extravagant. From a review of the observations of Tycho Brahe, Kepler, however, concluded that the error was actually in the opposite direction, and that the estimate must be raised to at least 13 million. In 1670 Cassini showed that these numbers were altogether inconsistent with the facts, and gave as his conclusion 85 million. The transit of Venus over the face of the sun June 3, 1869, had been foreseen and its great value in in the solution of this fundamental proposition in astronomy appreciated. With commendable alacrity various governments contributed their assistance in making observations, so that in Europe there were 50 stations, in Asia 6, in America 17."

"But on the discussion of the observation made at the various stations, it was found that THERE WAS NOT THE ACCORDANCE THAT COULD BE DESIRED—THE RESULT VARYING FROM 88 TO 109 MILLIONS. The celebrated mathematician, Encke, therefore revised them in 1822/4 and came to the conclusion that the sun's horizontal parallax, that is, the angle under which the semi-diameter of the earth IS SEEN FROM THE SUN, is 8.576/1000"; this gave as the distance 95,274,000 miles. Subsequently the observations were reconsidered by Hansen, WHO GAVE AS THEIR RESULT 91,659,000. Airy & Stone by another method, made it 91,400,000."

"Theoretical Astronomy" informs us to the following effect:

"Copernicus computed the distance of the sun from us to be 3,391,200 miles; Kepler reckoned it to be 12,376,800 miles; Riciola 27,360,000; Newton said it did not matter whether reckoned it 28 or 54 millions, for he said that *either* would do *well.* Benjamin Martin in his Introduction to the Newtonian Philosophy...says that its distance is between 81 and 82 millions of miles... Thomas Dilworth says 93,726,900 miles; Mr. Hind has stated positively that it is 95,298,260...Gillis & Gould say that it is more than 96 millions, and Mayer more than 104,000,000."

In the face of these alarming figures it would be wonder if astronomical enquiry were satisfied with approximate, or any other RESULTS, for results are just what cannot be arrived at.

Regiments of figures are paraded with all the learned jargon for which science is famous, but one might as well look at the changing clouds in the sky and seek for certainty there, as to expect to get it from the propounders of modern astronomy. The authoress of "Sun, Moon, and Stars," however, comes to the rescue of the learned and tells us that:

"It is only of late years that the matter has been *clearly settled.* And indeed, it was found quite lately that a mistake of nearly 3,000,000 miles had been made, notwithstanding all the care and all the attention given...the distance of the sun from the earth is *no less* than *about* 91,000,000 miles."

Following after a certainty in modern astronomy, is like following a phantom. Sir K. Ball, in his "Story of the Heavens," page 28, completely destroys this "clearly settled" matter, for he says (and he ought to know);

"The *actual* distance of the sun from the earth is about 92,700,00 miles."

That saving clause "*about*" is very handy indeed.

As the sun, according to "science" may be anything from three to one hundred and four million miles away, there is plenty of "space" to choose from. It is like the showman and the child. You pay your money — for various astronomical works — and you take your choice as to what distance you wish the sun to be. If you are a modest person, go in for a few millions: but if you wish to be "very scientific' and to be "mathematically certain" of your figures, then I advise you to make your choice somewhere *about* a hundred millions. You will at least have plenty of "space" to retreat into, should the next calculation be against the figures of your choice. You can always add a few millions to "keep up with the times," or take off as many as may be required to adjust the distance to the "very latest" *accurate* column of figures.

Talk about ridicule, the whole of modern astronomy is like a farcical comedy — full of surprises. One never knows what monstrous or ludicrous absurdity may come forth next. You must not apply the ordinary rules of common-sense to astronomical guesswork. No, the thing would fall to pieces if you did. But is there no means of testing these ever-changing never stable speculations and bringing them to the scrutiny of the hard logic of fact? Indeed there is. The distance of the sun can be measured with much precision, the same way as a tree or a house, or church steeple is measured, by plane triangulation. It is the principle on which a house is built, a table made or a man-of-war constructed It is used alike by the engineer and the carpenter, i Lot us put the statements of the learned as to the immense distance of the sun from the earth — anywhere between three and one hundred and four million miles — to this test.

When the sun is on the equator and thus has no declination, the angle it makes with the earth and sea at all points on that circle is a right angle. At an angular distance of 45° from the equator, north or south, the distance of the base line from the observer to the equator is of necessity the same as the sun's vertical distance from the earth's equator. That is to say, in any right-angled triangle where the angle at the apex of the triangle is 45°, the other angle must of necessity be the same; as these two angles in any such triangle are equal to the right angle, viz., 90. The angles being equal the sides are of necessity equal; therefore the base line is equal to the vertical. This principle holds good whether the triangle represents a field plotted by the surveyor; the measurement of the roof of a house erected by the builder; the distance a ship is from the land, known as the "four point bearing "; or the distance of a heavenly body measured with a sextant, the minutes and seconds of which correspond to miles and sixtieths of miles reckoned on the earth's surface. Whether the measurement is vertical as in the case of a housetop, church spire, or the sun in the heavens; or horizontal as in the case of the ship's distance from the shore, or the land plotted by the surveyor, the same principle holds good. It is the principle on which Cook measured the height of a tree, as the following quotation tells us. In "Cook's Voyages," by A. Kippis, page 54, it is said that:

"One of the trees at the height of six feet above the ground, was 19ft. 8in. in girt. Lieutenant Cook having a quadrant with him, measured its height from the root to the first branch, and found it to be 89 feet."

The following triangle illustrates this:

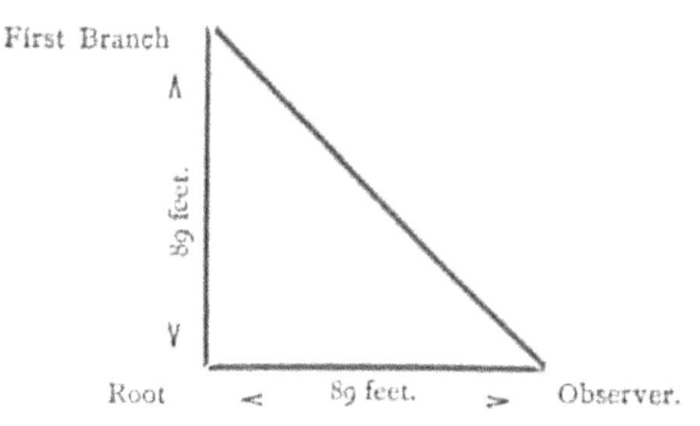

The reader will notice that the angle at the first branch is one of 45°, and the angle at the observer being the same, *the base line and vertical must be the same length* AND CANNOT BY ANY POSSIBILITY BE LESS OR MORE-Therefore if we can get a position 45° north or south of the equator when the sun has no declination, the distance from our place of observation to the equator (the base of the triangle), will be exactly equal to the distance of the sun from the earth's equator (the vertical).

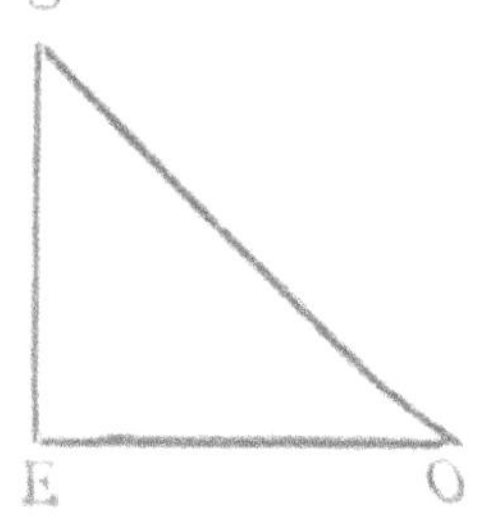

Let S E O be a right angled triangle, right angled at E; S the sun, E the equator, and O an observer at 45° north latitude.

From the figure it is evident that 45° is the angular distance of the sun at 45° north, and no other angle can be got in actual practice (allowing, of course, for such corrections as height of eye, semi diameter, &c.); so that the distance on. the surface of the earth to the equator—from O to E, is the same as from the equator to the sun in the heavens— E to S. Multiplying 45 by 60 (60 geographical miles = 1 degree), we get 2,700 geographical miles as the distance from O to E and thus from E to S.

THE SUN IS THERE-FORE 2,700 MILES DISTANT FROM THE EARTH. If the Sun were 96,000,000 miles distant from the Earth, an observer at 45° N or S latitude would be that distance from the Equator!!!

To make it perfectly clear to the *navigator,* let the following horizontal triangle represent the usual way the ship's distance from the shore is found, known as the four point bearing, to which reference has already been made:

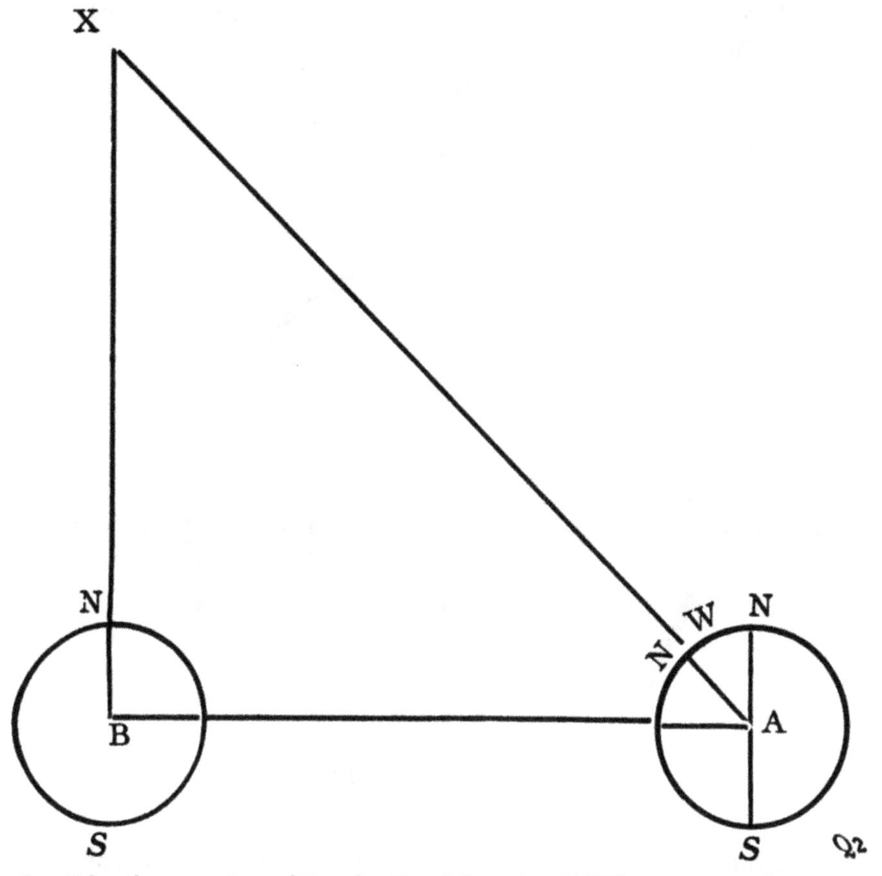

Let X be the position of Beachy Head, bearing N W by compass from a vessel bound down channel; A the position of the vessel when the headland bears N W, and B her position when the headland bears N by compass. It is required to determine the vessel's distance from Beachy Head, when the ship is at the position marked B. As the navigator will well understand, the vessel must be put on the course corresponding to the four point bearing, and as Beachy Head bears N W the course is West, and when the land is abeam and bears X, the distance the ship has sailed from the first position to the second one, is the same distance the ship is from the land at the point B.

If the navigator will apply this principle to the sun's distance, he will at once see that the distance of the sun from the earth cannot be either more or less than the distance of 45° of latitude from the equator, viz. 2,700 nautical miles.

It may be objected that this measurement is on the assumption that the waters of the world are horizontal. This I have produced abundant evidence to prove is the case, but even if the earth were the globe of astronomical imagination, the following diagram will show that the distance is in no wise altered, and would be the same if the observer could get an observation on a globular surface.

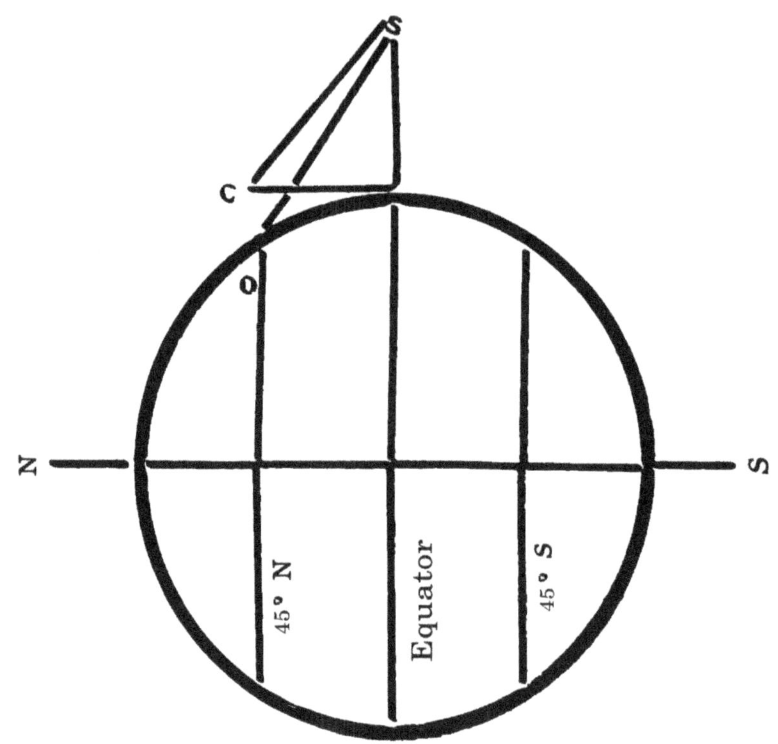

Let O be the place of observation at 45° north or south latitude, and S the sun when it has no declination; then the angular distance of the sun is less than 45°, *on account of the expression of the observer's position,* THEREFORE the angle O S C, must be added to the observation, being the allowance for CURVATURE to be made, which brings the observation to 45°. The distance on a globe, therefore, would be the same as on a flat surface, provided the observer could get an observation of the sun's angular distance on a globe, *which I have already shown to be impossible.* IT IS AS CERTAIN AS THAT TWO AND TWO ARE FOUR, THAT THE SUN'S DISTANCE FROM THE EARTH IS TWO THOUSAND SEVEN HUNDRED NAUTICAL MILES. We challenge the whole scientific world to disprove this statement.

Thirty-Three - Sun's Diameter

When we "read up" current science on the size of the sun, we shall find it as ridiculous and as far from the truth as the suns distance has been shown to be.

Sir Robert Ball, in his "Story of the Heavens," page 26, says that:

"The diameter of the orb of day...is 865,000 miles."

This is enlarged upon by R. Russell, who tells us that:

"The sun's diameter is 882,000 miles."

A. Giberne, in "Sun, Moon, and Stars," considerably lessens the value of the figures, for she tells us that:

"The diameter of the sun is *no less* than 850,000 miles."

Then G. F. Chambers, in his "Story of the Solar System," comes to the rescue with the true diameter and says:

"The TRUE diameter of the sun is 866,000 miles."

Let the reader observe that the *differences* of the sun's diameter, as given to us by professionals is no less than 32,000 miles, and let him decide as to which diameter he prefers.

The sun is always somewhere between the tropics of Cancer and Capricorn, a distance admitted to be less than 3,000 miles; how then can the sun if it be so many thousand miles in diameter, squeeze itself into a space of about 3,000 miles only? How can a locomotive seven feet wide run on a two feet gauge of rails? Can a camel ride on the back of a mouse, or a whale rush down the throat of a herring?

But look at the distance, say the professors. We have already done that and not one of the wise men we have so often challenged, has ever attempted to refute the principle on which we measure the sun's distance.

These tall figures of the sun's supposed diameter must be relegated to oblivion with as scant courtesy and as little ceremony as the sun's distance had to be thrown aside. Fact compels us to get rid of these absurd notions and to spread abroad the truth concerning them. What then is the diameter of the orb of day? Thirty-two miles, I reply. How is that obtained? By the same practical and non-theoretical manner as his distance was obtained. If the navigator neglects to apply the sun's semi-diameter to his observation at sea, he is 16 nautical miles (nearly) out in calculating the position his ship is in. A minute of arc on the sextant represents a nautical mile, and if the semi-diameter be 16 miles, the diameter is of course 32 miles. And as measured by the sextant, the sun's diameter is 32 minutes of arc, that is 32 nautical miles in diameter. Let him disprove this who can. If ever disproof is attempted, it will be a literary curiosity, well worth framing.

Thirty-Four - The Stars

In the "Story of the Heavens," to which I have so often referred, we find that the cardinal doctrine of astronomy is said to be:

"That the sun is no more than a star, and the stars no less than suns."

And on page 52 of the same work we are told that:

"Every one of the thousands of stars that can be seen with the unaided eye is enormously larger than our satellite."

In "A Treatise on Astronomy," by E. Henderson, it is noted that:

"The probability is that every star is a sun far surpassing in magnitude and splendour...Vega is 53,577 times larger than our sun."

The reader need not be alarmed at these statements, for *there is not one atom of truth in them,* THERE IS NOT A STAR IN THE SKY; NOT ONE BODY IN THE HEAVENS, THE SIZE OR DISTANCE OF WHICH IS KNOWN TO ASTRONOMERS. It is all speculation and guesswork,, but very poor speculation and miserably bad guesswork. They are wrong every time and always. The sun's distance is the datum for measuring the distances and sizes of all the heavenly bodies, and as it is hopelessly wrong, as we have shown, ALL THE SIZES AND DISTANCES OF ALL THE HEAVENLY BODIES ARE WRONG ALSO.

Thirty-Five - Star Distances

Sir Robert Ball, in his inimitable fairy tale already referred to (entitled the "Story of the Heavens"), says that:

"We now know the distances of a few of the stars, perhaps 20 or 30, with more or less accuracy, but of the distances of the great majority we are still ignorant...The observations for the determination of stellar parallax *are founded on the familiar truth that the earth revolves around the sun.*"

The statement, that "we now know the distances" is unconditionally false. They do not know any one distance. Neither can they know, because the speculation is founded on a myth — the earth's supposed revolution round the sun, which I have shown to be impossible. But let us proceed, and see with what marvellous "accuracy" the distances are known.

On pages 414 to 421 of the work referred to, we find that:

"Bessel concluded that the distance (61 Cygni) was about 60 billion miles. Struve thought it could not be more than 40 billions of miles."

A *little* difference of 20,000,000,000,000 miles. How very *accurate,* to be sure.

Sir Robert then calmly informs us that:

"We shall presently show that we believe Struve was right, *yet it does not necessarily follow that Bessel was wrong.*"

What splendid logic, and what marvellous reasoning faculties! He then continues:

"As the distance of 61 Cygni is 40 billions of miles."

So that after all the putting forth of mighty intellectual power it seems that Bessel was wrong, because Sir Robert says that the star is 43 billion miles away, which is the distance given by Struve. And then follows an audacious statement:

"By the aid of our KNOWLEDGE OF STAR DISTANCES, combined with an *assumed* velocity of 30 miles per second, we can make the attempt to peer back into the remote past."

No, Sir Robert, you have not yet shown that you know the slightest of the *present* in your own profession, so we cannot take you as a guide to enable us to "peer back" into the past.

But how are star distances measured? Mr. Laing shall tell us. In his "Modern Science and Modern Thought," page 8, he says:

"The distance of the earth from the sun being 93 million miles, and its orbit an ellipse nearly circular; it follows that in mid-winter, in round numbers, it is 186 million miles distant from the spot where it was at midsummer."

This is all supposition, which I have already shown does not contain a word of truth, and consequently whatever is built upon this foundation is worthless. Now it is evident to every thinking man that if the earth has travelled such an enormous distance in an ellipse so as to make the baseline 186 million miles, all the stars will necessarily have altered in relative position, so that the matter can be easily tested. Now, what says Mr. Laing:

"What difference in the bearing of the fixed stars is caused by traversing this enormous base? The answer is, *in the immense majority of cases* NO DIFFERENCE AT ALL."

In the time of Tycho Brahe it was said that the earth revolved around the sun, but he argued that if the earth revolved around the sun, the relative position of the stars would change very much, and the matter must, in the nature of the case, be easily detected. Accordingly experiments were tried at intervals of six months, and the result showed that the stars were in exactly the same position as they had occupied six months before, thus proving that the earth does not move at all. The "explanation" Mr. Laing gives is nullified by his own further statement. He tells us that:

"Their distance is so vastly greater than 186 million miles, that a change of basis to this extent makes no change perceptible to the most refined instruments in their bearings as from the earth."

The distance of the stars is an absolutely unknown quantity to the gentlemen of the observatories, as I have shown, so that this flimsy argument amounts to nothing. Besides this, the movement of the earth, if such ever took place, would be easily detected. But that such has never been observed, and that the relative position of the stars has not changed, proves that the earth is a fixture.

Mr. Laing goes on to refute his own statement of the case by stating that:

"*The perfection of modern instruments is such,* that A CHANGE OF EVEN ONE SECOND, OR ONE-THREE-THOUSAND-SIX-HUNDREDTH PART OF ONE DEGREE, in the annual parallax, as it is called, of any fixed star, WOULD CERTAINLY BE DETECTED."

By the most powerful and finely adjusted of modern instruments no change has ever been observed, so that Mr. Laing's laboured statement must be relegated to the limbo of conjectural absurdity.

Mr. Laing's case against the Bible would be the most telling that could be made out, if his statements were within a million miles of the truth, but they are absolutely without the slightest foundation and must be thrown into the "scientific" waste-paper basket.

Another writer who uses his not inconsiderable ability in the same direction is Dr. Draper, author of a work I have already quoted from, "The History of the Conflict between Religion and Science." On the subject of star distances, he says, page 156:

"Considering that the movement of the earth does not sensibly affect the apparent position of the stars, he (Aristarchus *inferred* that they are incomparably more distant from us than the sun) ...He saw that the earth is of absolutely insignificant size when compared with the stellar universe. *He saw too, that there is nothing above us but space and stars.*"

What a marvellous vision this man must have had! Had it only been stated what Planet this adventurer chartered to take his trip "above us" to see what there was there, the fairy tale would have been complete.

Thirty-Six - The Seasons

R. Russell tells us in his "Wonders of the Sun, Moon and Stars," pages 16 and 17, that:

"The nearer the sun gets to the Pole Star the earlier it is, the higher it reaches at noon, and the later it sets; and the further it gets from the Pole Star the later it rises, the lower it is at noon, and the earlier it sets. This *apparently independent motion of the sun* therefore, seems to account for longer and shorter days and the whole phenomena of the seasons; but why the sun lags as described, or why it moves northerly and southerly at alternate periods, *there is no apparent evidence.*"

On the supposition that the world is a globe rotating against the sun, and revolving round that luminary, it is impossible to account for what Mr, Russell calls the lagging movement of the sun. But on a flat surface like the world is known to be, there is no assumption needed to account for it. As I have shown, the earth is a stretched-out structure, which diverges from the central north in all directions towards the south. The equator, being midway between the north centre and the southern circumference, divides the course of the sun into north and south declination. The longest circle round the world which the sun makes, is when it has reached its greatest southern declination. Gradually going northwards the circle is contracted. In about three months after the southern extremity of its path has been reached, the sun makes a circle round the equator. Still pursuing a northerly course as it goes round and above the world, in another three months the greatest northern declination is reached, when the sun again begins to go towards the south. In north latitudes, when the sun if going north, it rises earlier each day, is higher at noon and sets later; while in southern latitudes at the same time, the sun as a matter of course rises later, reaches a lesser altitude at noon and sets earlier. In northern latitudes during the southern summer, say from September to December, the sun rises later each day, is lower at noon and sets earlier; while in the south he rises earlier, reaches a higher altitude at noon, and sets later each day. This movement round the earth daily is the cause of the alternations of day and night, while his northerly and southerly courses produce the seasons. When the sun is south of the equator it is summer in the south and winter in the north; and *vice versa.* The fact of the alternation of the seasons flatly contradicts the *Newtonian* delusion that the earth revolves in an orbit round the sun. It is said that summer is caused by the earth being nearest the sun, and winter by its being farthest from the sun. But if the reader will follow the argument in any text book he will see that according to the theory, when the earth is nearest the sun there must be summer in both northern and southern latitudes; and in like manner when it is farthest from the sun, it must be winter all over the earth at the same time, because the whole of the globe-earth would then be farthest from the sun!!! In short, it is impossible to account for the recurrence of the seasons on the assumption that the earth is globular and that it revolves in an orbit round the sun.

Thirty-Seven - Signals on Sea and Land

Pearson's Weekly of the 29th December, 1894, says:

"Evidently we have not got at the bottom of the matter yet. In August, 1890, the C Manouvre Fleet signalled with searchlights to Colliers, 70 miles away...The information comes from Mr. F. T. Jane, the Artist who was on board at the time."

According to the Astronomers, these vessels should have been 3,200 feet below the horizon, allowing for a height of 40 feet on the signalling vessel, and 26 feet on the Colliers!!!

Harper's Weekly of 20th October, 1894, contains particulars of an experiment made by the Signal Corps of the U.S. Army, with the Glassford flashlight or heliograph.

The signal stations were Mount Uncompahgre, in South Western Colorado, and Mount Ellen in Southern Utah; the former 14,418 feet above sea level, the latter 11,410 feet; the plateau lying between the two stations is 7,000 feet higher than the sea. According to the calculated rate of curvation of a spherical body of 25,000 miles in circumference, a straight line running at *right angles* with the *perpendicular* at the transmitting station. Mount Uncompahgre, would run as a *tangent* from the line of curvation so that in the distance of 183 miles, the curvation would place Mount Ellen *downward* from the tangent line, *below* the line of vision *nearly* 3¾ *miles!* and yet the receiving station was seen on *a line with the eye* from Mount Uncompahgre, on a line *coincident with the "tangent" line!!!*

Thirty-Eight - Surveying

In Robinson's "New Navigation and Surveying," page 25, it is stated:

"The spirit-level...is used to determine a horizontal line. A horizontal line is at right angles to the vertical. It is a level line."

And on page 33 the following occurs:

"To adjust a theodolite measure very carefully the distance between two stations, and set the instrument half way between them. Now bring the level near to one of the stations, level it carefully and sight the rod. Note the number on the rod, say 6 feet, and have the rod man go to the other station and place his target on the rod, just 6 feet. When the telescope is turned upon it the horizontal spider line ought to just coincide with the target, and will if the instrument is level or in perfect adjustment."

This proves that the whole of the line from the extremities at either side of the instrument, passing through the telescope is a level or straight line, impossible on a globe. And the further fact that in surveying, no allowance is made for the supposed curvature of the earth, demonstrates that the earth is a plane. The surveyor is, in many cases, deluded by the speculations of the learned. They tell him that because he takes his sights midway between two stations, the allowance for curvature is made. But we have shown from a text-book that the line is a level or straight line, so that the learned are all wrong. And if a section of a globe be drawn and the instrument shown at various equal distances, *to get a continuous straight line the instrument would have to be taken up off the globe into space.*

That in all surveys no allowance is made for curvature, which would be a necessity on a globe; that a horizontal line is in every case the datum line, the same line being continuous throughout the whole length of the work; and that the theodolite cuts a line at equal altitudes on either side of it, which altitude is the same as that of the instrument, clearly proves, to those who will accept proof when it is furnished, that the world is a plane and not a globe.

Thirty-Nine - Science

"Lux" of the 13th January, 1894, has the following:

"What a lovely thing the word 'science' is! There was an old lady who, in times of trouble and anxiety, always found comfort and peace in that blessed word, Mesopotamia. But that aged person is not in it with the old women who find a solace in that blessed word 'science'. The latest thing in 'science' is the 'Interstellar Medium'. Space is not void, we are to believe as commanded by 'science', but it is filled with a kind of stuff called ether. It conveys lights from the stars at, say, the rate of 186,300 miles per second. Light comes in waves. The waves have a mean value of 50,000 to the inch. This light comes 60,000,000,000,000,000 waves in one second of time. Some stars, according to Herschel, take 300,000 years to send their light to our earth. Go on, work it out! When found, make a note of it, and then say 'science' doesn't want about 1,000 times more faith than Christianity, if you can!"

In "Paul Petoff," by F. Marion Crawford, on page 117 it is stated:

"We talk more nonsense about science than would fill many volumes: because, though we devote so much time to the pursuit of knowledge, nevertheless the amount of knowledge actually acquired, beyond all possibility of contradiction, is ludicrously small as compared with the energy expended in the pursuit of it, and the noise made over its attainment. Science lays many eggs, but few are hatched. Science boasts much, but accomplishes little; is vainglorious, puffed up, and uncharitable; desires to be considered the root of all civilization, and the seed of all good, whereas it is the heart that civilises, and never the head."

"Sigma," in the "English Mechanic" for 5th October, 1894, supplements the above as follows:

"We have any quantity of hypotheses thrust upon us as discoveries, which are merely false knowledge that later science will have to unlearn. As a matter of fact, the fashionable notions that are paraded as 'science' stand only because their advocates shut their eyes to realities, make assertions with little or no fact to start from, ignore the facts which do not suit them, refuse to meet objections, and ignore any really scientific (that is provable) explanations which do not agree with the specialistic facts."

"Science" is a very inclusive term, as the foregoing extracts show. It is the cloak under which thousands of humbugs flourish and grow great, "science," however, sometimes exposes "science," as the following from "Modern Science and Modern Thought," page 43, shows:

"In this state of things the moon is supposed to have been thrown off from the earth...Now these conclusions may be true or not as regards phases of the earth's life prior to the Silurian period, *from which downwards* GEOLOGY SHOWS UNMISTAKABLY THAT NOTHING OF THE SORT, OR IN THE LEAST DEGREE APPROACHING IT, HAS OCCURRED."

When Geology mocks at Astronomy two combatants to fight it out, for they are both fables.

The "English Mechanic" of 4th January, 1889, says:

"The whole of astronomical science so far as the stellar universe is concerned is founded upon a false basis. This arises from the fact that the construction of the heavens in respect to the apparent arrangements of the stars in space is always erroneous, and yet necessarily all astronomy is founded upon this suppositious situation of the stars."

Commenting on "Scientific Dogmatism," the "Daily News" of 5th December, 1893, says:

"Mr. Tyndall resigned in 1887 the Professorship at the Royal Institution which he had held for more than thirty years...He never had any doubt about anything, from Home Rule to spontaneous generation, from the composition of dust to the origin of things...But while Professor Tyndall, the brilliant lecturer, the luminous expositor, the intrepid climber, the pugnacious controversialist, the genial and amiable companion, was in many respects an interesting personage, no part of his character would repay study so well as the scientific dogmatism in which it was all steeped. Dr. Arnold protested half a century ago in his entertaining, if not very practical, notes on Thucydides, against what, as a philological student, he discerned to be a tendency of the times. 'It is not to be endured, he said, that scepticism should run at once into dogmatism, and that we should be required to doubt with as little discrimination as we were formerly called upon to believe.' Dr. Arnold was of course referring directly and immediately to the tampering of commentators with the text of the Greek historian. But the symptom which he observed has spread into other spheres, and for the old tyranny of the Church there has been substituted the despotism of the laboratory. The 'delight of dealing with certainties' described by an accomplished man of letters, who made an hasty plunge into the 'Principia', is a high form of mental enjoyment. But it is rather a dangerous guide through the maze of conflicting probabilities, from which even the sacred College of Science has not yet succeeded in delivering the human race...

Mr. Balfour wrote a book which is not nearly so well-known as it ought to be. The 'Defence of Philosophic Doubt' is dry and unattractive in form. But it is acute and ingenious in substance. It would be a more agreeable work if it were written in literary English. It would be a more candid one if it mentioned the name of David Hume. It is, notwithstanding these drawbacks, a value-able antidote to the pretensions of modern science. In it Mr. Balfour, one of the few living Englishmen with a real aptitude for philosophy, turns against the exaggerated claims of science the argument formerly employed with so much vigour against the exaggerated claims of theology. 'It is useless,' he says in effect, 'to tell me that your conclusions are true because they are universally accepted. What is the ignorant impression of the unthinking multitude really worth?'...Mr. Balfour is fond of paradox, and he may press his theory too far. But at least he deserves credit for pointing out that the infallibility of science rests on no sound foundation than any other form of orthodox opinion. The greatest names in scientific history cannot be cited to support the doctrine that a knowledge of physics, however accurate and extensive, entitles its possessor to lay down the law on final causes and the origin of things. In his famous address at Belfast nearly twenty years ago, Professor Tyndall declared that matter contained the power and potency of every form of life. If this phrase was more than empty rhetoric it implied that Professor Tyndall knew how the world came into existence, and how life began. Mr. Darwin, the greatest man of science since Newton, if not since Aristotle, put forward no such assumption. In humble and dignified language he explained that his marvellous generalisations with reference to the origin of species and the decent of man began, as they ended, with a living creature. He traced man to the marine ascidian. The marine ascidian he did not pretend to trace."

Forty - The Tides

It is commonly taught that the tides are caused by lunar attraction. Sir Robert Ball tells us that:

"The moon attracts the solid body of the earth with greater intensity than it attracts the water at the other side which lies more distant from it. The earth is thus drawn away from the water, which accordingly exhibits a high tide as well on the side of the earth away from the moon as on that toward the moon. The low tides occupy the intermediate positions."

No one who has the use of all his faculties and who dares to use them, need be told that this flimsy apology for what the learned cannot account for, contradicts itself. How could this attraction take place without disintegrating the globe? Besides, as the law of gravitation is said to operate according to the amount of matter of which each body consists, the statements of astronomers that the moon is 2,160 miles in diameter and the earth 8,000 miles in diameter flatly contradict their own other statements about the moon causing tides. How can the smaller body attract the larger? We are informed in "Sun, Moon, and Stars," pages 160 to 163, that:

"The earth, it is true, attracts the moon. So also the moon attracts the earth; THOUGH THE FAR GREATER WEIGHT OF THE EARTH MAKES HER ATTRACTION TO BE FAR GREATER."

How anyone can accept the current theory in face of the above is somewhat puzzling. Sir R. Ball says the moon attracts the solid body of the earth; but the work from which I have just quoted states that:

"Her attraction (the moon's) draws up the yielding waters of the ocean in a vast wave."

Both these assertions cannot be true. Which is? I say neither. And the astronomers' own theory of attraction also answers "neither" when it is taken into consideration that the moon cannot attract the earth, being a much smaller body.

But if the moon lifted up the waters, it is evident that near the land, the water would be drawn away and *low*, instead of high tide, caused. Again, the velocity and path of the moon are uniform, and it follows that if she exerted any influence on the earth, that influence could only be a uniform influence. But the tides are not uniform. At Port Natal the rise and fall is about six feet, while at Beira, about 600 miles up the coast, the rise and fall is 26 feet. This effectually settles the matter that the moon has no influence on the tides.

How then are tides caused? The learned being as far from the truth in this as in every matter which we have brought to the test of the hard logic of facts, what is the truth of the matter?

The Leicester Daily Post of 25th August, 1892, says:

"M. Bouquet de la Grye, an eminent hydrographical Engineer, has after long years of study calculated the atmospheric expansions and depressions which coincide with spring and neap tides. There have been cases in which air was moved in waves of 133 yards high, and in places where the barometrical pressure was seven-tenths of an inch, of six and a half miles. Near the upper surface of the earth's atmosphere condensations and dilations of this magnitude are frequent. The human nervous system may be said to register these air waves. We are only aware that they do so by the discomfort which we feel. The earth also registers them and to its very centre. The incandescent and fluid matter under the earth's crust acts in concert with the air and sea at the full of the moon. In 1889 a German Scientist, Dr. Rebeur Pachwitz, thought he noticed at Wilhelmshaven and Potsdam earth oscillations corresponding with the course of the moon. He wrote to the observatory at Tenerife asking for observations to be ma.de there in December, 1890 and April, 1891, which would be propitious times for them. *From these observations and others simultaneously made in the sandy plains round Berlin*, IT WAS ESTABLISHED THAT THE Earth RISES AND FALLS LIKE THE OCEAN OR THE ATMOSPHERE. The movements, common to them all, may be likened to the chest in breathing.—Paris Correspondent Weekly Dispatch."

This is the answer to the question. Tides are caused by the gentle and gradual rise and fall of the earth on the bosom of the mighty deep. In inland lakes, there are no tides; which also proves that the moon cannot attract either the earth or water to cause tides. But the fact that the basin of the lake is on the earth which rests on the waters of the deep, shows that no tides are possible, as the waters of the lakes together with the earth rise and fall, and thus the tides at the coast are caused; while there are no tides on waters unconnected with the sea.

The "Yellow Frigate," by Jas. Grant, page 189, states:

"St. Mungo's Tide. This double flow is somewhat remarkable, for when the tide appears full it suddenly falls fifteen inches, and then returns with greater force, until it attains a much higher mark."

The following is from "Omoo, a Narrative of Adventures in the South Seas," by H. Melville:

"The Newtonian theory of the Tides does not hold good at Tahiti, where, throughout the year, the waters uniformly commence ebbing at noon and midnight, and flow about sunset and daybreak. Hence the term 'Toorerar-Po' is used alike to express highwater and midnight."

The question may now be asked, what has the moon to do with the tides? The moon is the TIMEKEEPER for the tides, nothing more. The "phase" of the moon tells what kind of a tide may be expected, but she does not and cannot "attract" either the solid body of the earth or the waters. What Zetetics have stated for many years past, is now seen to be true, but "science" is slow to take advantage of the fact.

Forty-One - The Ultimate Conclusions of Science in Relation to Bible Teaching

In the preceding pages it has been clearly shown that the Copernican or Newtonian System of Astronomy is an absurd composition of meaningless expressions, false ideas, and mechanical impossibilities. In our consideration of the subject — and we have touched upon all the important items — we have not found one statement which does not require a supposition to start with; not a single fact has been elicited from the published books on the subject written by the profession; and contradictions have been found in all the most important component parts of the "science," which effectually refute the system and destroy its claims. Hence, the whole hypothesis must be rejected as a snare and a delusion, without a vestige of fact or possibility to support its bold, unwarranted, and Infidel conclusions.

I shall now proceed to demonstrate that when the fictions of the system are received as facts, the logical necessity arises for disposing of the Bible as a collection of old wives' fables. I shall also quote from the Scriptures J themselves, to prove conclusively that NATURE and the BIBLE *are in perfect agreement.*

In Paine's "Age of Reason," it is stated that;

"The two beliefs—modern astronomy and the Bible—cannot be held together in the same mind; he who *thinks* he believes both has *thought* very little of either."

However much many well-meaning Christians may affect to ignore this statement, it is nevertheless true. The system of astronomy at present in vogue is the very opposite of the facts of nature, as we have abundantly demonstrated. The facts of nature are in perfect harmony with the Bible, as we shall presently see.

The most casual and superficial reader of the Bible must see that it claims to be of Divine Origin. He must further see that the Author of the Bible claims to be the Builder of the Universe. And he must still further see that the world is described in this Book which claims to be from God as being built upon the waters of the mighty deep, which foundations are not to be discovered by man j that the Sun, Moon, and Stars are inferior to the world we live on, and that they move above the earth, which is at rest.

How, then, can a thinking person affect to believe the Bible *and* a system which teaches the very opposite of the teaching of that Volume. The logical conclusion is that if the statements of modern astronomy be true, the Bible cannot be what it claims to be — THE WORD OF GOD. We have already shown that there is not so much as one true statement in all modern astronomy concerning this world—that the whole thing is a fake and a fable, an ingenious hoax. It is, therefore, not incumbent on any one to believe the imposture; but all lovers of truth should join hands in exposing the thing. We shall now see that the extravagant and false ideas of the scientific world have led the more daring intellects to despise Bible teaching, and, in some cases, to reject the idea of the existence of a personal God at all. But we shall also show that such conclusions are merely the logical sequence of belief in the impossible theories of the "learned." Two opposite things cannot both be true, and the "scientists" thinking that modern astronomy is true, have only been acting logical manner by rejecting the teaching of the Bible.

R. A. Proctor, in his work entitled "Our Place Amongst Infinities" page 3, unblushingly states:

"To speak in plain terms, *as far as science is concerned,* THE IDEA OF A PERSONAL GOD IS INCONCEIVABLE, at are also all the attributes which religion recognises in such a being."

A Durban gentleman told the writer some time that:

"When the Bible speaks of physical things such as the earth, IT IS ABSOLUTELY UNTRUE."

And a "reverend" gentleman told me in April, 1898, that:

"The Bible is only inspired when it speaks on matters of the soul; when it speaks on physical matters, such as astronomical facts, IT IS MERELY THE OPINION OF THE WRITERS."

But if the first two statements are only the logical sequence of believing the fictions of modern science to be facts, what shall we say about the third? It is much more inconsistent than anything that the avowed enemies of the Gospel could devise. They believe science and *therefore* disbelieve the Bible, which is contrary to science. But to believe *both* to be correct as some do, or to say that when the Bible speaks of physical facts if is only the opinions of the writers and not inspired, is to refute any statement made as to inspiration in any other direction.

Obviously, if the Bible be not true in matters scientific it cannot possibly be true on any other matter. It is either true in part and true altogether, or false in part and false altogether. Between modern astronomy and the Bible, there is not so much as an inch of standing ground; if the one be true the other and opposite statement is false.

But there are a great many Christians who do not seem able to arrive at any logical conclusion in the matter. They take for granted that what science teaches is true, because many "learned" men believe it. But when brought face to face with the fact that Bible and astronomical teaching are contrary the one to the other, and because men believe science, *therefore* they disbelieve the Bible; they at once begin to say that the statements in the Bible concerning the world are merely "poetic" or "symbolic" and by no means literal, But before arriving at such a conclusion it must, in all fairness, be shown that those passages which teach that the world is at rest, and the sun, moon, and stars are moving over and around it, are consistent with other passages which are, admittedly, not symbolical, but literal beyond all controversy. I may instance Joshua commanding the sun to stand still, which, if the reference to its movement in Psalm 19 be symbolical and not literal, brings to light a, serious discrepancy, for the Scriptures say that the sun did stand still. Now, according to modern astronomy, the sun never does anything but stand still. Does it not, therefore, seem very absurd that a General of a large army should be so ignorant about such a simple matter, of which his God had already spoken, and yet be the leader of a people called out of Egypt by God; not knowing whether the sun or the world moved; and must not the Scripture which distinctly stales that the sun was made to stand still, be very absurd, if the sun always stands still?

Then again, Christ is said to have been shown all the kingdoms of the world in a moment of time. This is admittedly literal. But if the passages which refer to the world standing still be symbolical, and the world be moving, turning upside down in fact, it would have been quite impossible for Christ to have seen all the kingdoms of the world in a moment, as some of them would be far below the horizon, on the other side of the revolving ball.

Many such statements could be produced, showing the absurdity of the symbolical idea, and clearly indicating that the question in its literalness must be faced, whatever the issues be.

If the Christian thinks that the Scriptures are symbolical in this matter, the infidel, who searches the volume in order to find discrepancies, knows that it is very literal; and com. paring one passage with another very soon discovers that, from Genesis to Revelation, there is a marvellous consistency of teaching that the world is at rest and that sun, moon, and stars move around and above it- He therefore concludes that, inasmuch as Bible teaching is opposed to what he is pleased to denominate the "ascertained facts of science," the Bible must be untrue in matters scientific, AND THEREFORE, UNTRUE IN EVERY PARTICULAR. and if the reader will just apply the ordinary rules of common-sense, he will see that if the Bible be not true in some things, it cannot be true in any, and, therefore, must be rejected *in toto*. If, for example, the world be the globe in popular belief, it is impossible that there ever could have been a universal flood. For such a thing to have happened, it would be required to blot out the whole universe, to stop the revolution of the globe and to bring confusion and ruin to the whole of the "solar system." But the Bible does such that there was a universal deluge, and that is admittedly literal. Not only so, but Christ refers to the deluge. If, therefore, no deluge ever happened, it would be very inconsistent to ask any one to believe in Christ, who testified that that great catastrophe actually took place. In our present enquiry, therefore, we must leave the whims and prejudices of those who say they believe the Bible, and yet except as truth the teaching of modern astronomy, which is the direct opposite of, and gives the lie to, Bible teaching; and see where the acceptation of the globular theory has led men to. If it were consistent with Bible teaching, it would naturally lead them to the Bible and the Christ of the Bible; inconsistent with the facts of the Bible, it could only lead men to doubt and deny that Book.

In *Lucifer,* of 23rd December, E.M. 287 (*i.e.* 1887 A.D.), the following occurs:

"We date from the first of January, 1601, This era is called the Era of Man (E. M.) to distinguish it from the theological epoch that preceded it. In that epoch the earth was supposed to be flat, the sun was its attendent light revolving about it. Above was Heaven, where God ruled supreme over all potentates and powers, is the kingdom of the Devil, Hell. So taught the Bible. Then came the NEW ASTRONOMY. It demonstrated that the a globe revolving about the sun; that the stars are worlds and suns; that there is no 'up and down' in space, VANISHED THE OLD HEAVEN, VANISHED THE OLD HELL; the earth became the home of man. And when the modern cosmogony lame, the Bible and the Church as infallible oracles had to go, for they had taught that regarding the universe WHICH WAS NOW SHOWN TO BE UNTRUE IN EVERY PARTICULAR."

In *Reynolds' Newspaper*, of 14th August, 1892, under the heading of "Democratic World," the following appeared:

"We are trembling on the eve of a discovery which may revolutionise the whole thought of the world. The almost universal opinion of scientific men is that the Planet Mars is inhabited by beings, like or superior to ourselves. Already they have discovered eat canals cut on its surface in geometrical form, which can only be the work of reasoning creatures. They have seen its snowfields, and it only requires a telescope a little stronger than those already In existence to reveal the mystery as to whether sentient beings exist on that planet. IF it be found that this is the case, THE WHOLE CHRISTIAN RELIGION WILL CRUMBLE TO PIECES THE STORY OF THE CREATION HAS ALREADY BECOME AN OLD WIFE'S TALE. HELL IS NEVER MENTIONED IN ANY WELL-INFORMED SOCIETY OF CLERGYMAN; the devil has become a myth. IF Mars is inhabited, the irresistible deduction will be that all the other planets are inhabited. This will put an end to the fable prompted by the vanity of humanity that the Son of God came to earth and suffered for creatures WHO ARE THE LINEAL DESCENDENTS OF MONKEYS. It is not to be supposed that the Hebrew carpenter, Jesus, went about as a kind of theosophical missionary to all the planets in the solar system, reincarnate, and suffering for the sins of various pigmies or giants, as the case may be, who may dwell there. The astronomers would do well to make haste to reveal to us the magnificent secret which the world impatiently awaits."

Professor W. B. Carpenter, in his paper in the *Modern Review* for October, 1880, protests that science has excluded God from Nature. He says;

"While, however, the idea of Government by a God IS NOW EXCLUDED BY GENERAL CONSENT FROM THE DOMAIN OF SCIENCE, the notion of Government by law has taken its place, not only in popular thought, but in the minds of many who claim the right to lead it; and it is the validity of this notion which I have now to call in question...PHILOSOPHY FINDING NO GOD IN NATURE NOR SEEING THE WANT OF ANY."

"The advanced philosophy of the present times goes still farther, asserting that THERE IS NO ROOM FOR A GOD IN NATURE."

These conclusions are the inevitable result of believing the current theories regarding the evolution of the world in opposition to Bible statements, that it is the product, not of evolution, but of special creation. This is the conclusion to which the world is fast hastening — NO ROOM FOR GOD IN NATURE. And when natural truth is rejected to keep pace with unnatural and fictitious science, no marvel if spiritual truths as revealed to man by his Creator, are rejected also. The one is the natural outcome of the other, S. Laing, in his "Modern Science and Modern Thought," tells us that:

"Attempts to harmonise the Gospels and prove the inspiration of writings which contain manifest errors and contradictions, have gone the way of Buckland's proof of a universal deluge, and of Hugh Miller's attempt to reconcile Noah's Ark and the Genesis account of creation WITH THE FACTS OF GEOLOGY AND ASTROMOMY."

The words "the facts of geology and astronomy" reveal the whole of the case for the infidel. He supposes that his assumptions are true. He assumes that his assertions are facts and THEREFORE the Bible, which tells against his so-called "facts" must be untrue.

I have already shown that astronomy has not yet chronicled one fact regarding this world; that the "facts" of astronomy regarding the enormous size, and by consequence the immense distance of the stars, are fictitious every one; that, in fact, modern astronomical "science" untrue altogether and unworthy the credence of any man, therefore the GREAT OUTCRY made by the "scientific" world against the Bible has ABSOLUTELY NO FOUNDATION.

On pages 178 and 179 of Draper's "Religion and Science," it is said:

"In his 'Evening Conversations' he (Giordano Bruno) had insisted that the Scriptures were never intended to teach science, but morals only; and that they cannot be received as of any authority on astronomical and physical subjects. Especially must we reject the view they reveal to us of the constitution of the world, that the earth is a flat surface, supported on pillars: that the sky is a firmament — the floor of heaven. On the contrary we must believe that the universe is infinite, and that it is filled with self-luminous and opaque worlds, many of them inhabited."

Bruno, like many now, was afraid of incurring the wrath of the priesthood by stating that the Bible was untrue, so he made a kind of compromise, as the above extract shows. But his argument does not require a second reading to show that if the science of the Bible be untrue, its moral teaching it must be equally so. Mr. Laing further tells us:

"Now it is absolutely certain that portions of the Bible, and these important portions relating to the creation of the world and of men *are not true and therefore* not inspired. IT IS CERTAIN THAT THE SUN, MOON, STARS AND EARTH WERE NOT CREATED AS THE AUTHOR OF GENESIS SUPPOSED THEM TO HAVE BEEN CREATED...IT IS CERTAIN THAT NO UNIVERSAL DELUGE EVER TOOK PLACE SINCE MAN EXISTED."

on pages 278 and 279 he adds:

"It is as certain as that two and two make four, THAT THE WORLD WAS NOT CREATED IN THE MANNER DESCRIBED IN GENESIS; THAT THE SUN. MOON AND STARS ARE NOT LIGHTS PLACED IN THE FIRMAMENT OR SOLID CRYSTAL VAULT OF HEAVEN, TO GIVE LIGHT UPON THE EARTH..."

This "absolute certainly" is the creation of the imagination, for there is not one FACT in nature that modern science can bring forward in support of the contention. The whole thing, from start to finish, is a myth, as we have abundantly demonstrated, and must be rejected.

Mr. Laing further says that:

"The conclusions of science are irresistible, and old forms of faith, however venerable and however endeared by a thousand associations, have no more chance in a collision with science than George Stephenson's cow had, if it stood on the rails and tried to stop the progress of a locomotive."

From purely practical data we have already seen that "the conclusions of science" are as unreasonable and fallacious as it is possible for the human mind to conceive. A mixture of infidel superstitions and gross absurdities constitute the most of present-day science respecting the world we live on. Its relation to truth is as darkness to tight. Science has as much chance in a collision with TRUTH as a rotten ship would have in a collision with an iron-clad.

Even professedly Christian people are hoodwinked and by modern hypothetical science.

Giberne in "Sun, Moon and Stars," says, when speaking of the Moon:

"All is dead, motionless, still. Is this verily a blasted world? Has it fallen under the breath of Almighty wrath, coming on scorched and seared?"

The "lesser light" that God declares He made to "rule the night" is set down as a blasted world, and that by a professed Christian! To this end the teaching of modern astronomy tends to "attract" all who receive its dicta, and cannot, therefore, be retained in the same mind with the Bible.

A noteworthy feature of the present day is the fact that many so-called Christian ministers are joining hands with the enemies of the Bible to teach the people that the Old Book is so very unscientific that it can no longer be regarded in the light of a word from God at all.

In the "*Christian World Pulpit*," of 14th June, 1893, the Rev. C. F. Aked is reported as saying, at Pembroke Chapel, Liverpool, that:

"No student of science is able to believe that any such flood as that recorded in the early chapters of Genesis ever took place in the history of the human race...The Flood story IS A MYTH, 'not history.'"

This gentleman has arrived at this conclusion by supposing that science is truth, and he is logically forced to believe that the Bible is a myth. Then what say the avowed enemies of the Book of God? Says the *Freethinker*, of 16th October, 1892:

"There is something in Christianity calculated to make it hostile to science. Its sacred books are defaced by a puerile cosmogony, and a vast number of physical absurdities; while its whole atmosphere, in the New as well as in the Old Testament, is in the highest degree unscientific.

The Bible gives a false account of the origin of the world; a foolish account of the origin of man; a ridiculous account of the origin of languages. It tells us of a universal flood which never happened. And all these falsities are bound up with essential doctrines, such as the fall of man and the atonement of Christ; with im-

portant moral teachings and social regulations. It was therefore inevitable that the Church, deeming itself the divinely-appointed guardian of Revelation, should oppose such sciences as astronomy, geology, and biology, which could not add to the authority of the Scripture, but might very easily weaken it. Falsehood was in possession, and truth was in exile or a prisoner."

This is clinched by the Public Press which teaches people to think. *Reynolds' Newspaper,* of 13th October, 1895, says:

"The most noteworthy feature of the British Association this year is that the assembled *savants* — representing *religion,* science, philosophy, politics — *have surrendered hands down* to views which, if accepted by anyone ten years ago, would be sneered at as a mark of disgrace. The Church has had to give in because *geology* and *biology* have been *too strong* for the *Book of Genesis,* which is no longer to be accepted as a real account of the Creation, but merely a symbolical one. The *incontestable experiments* and experiences of the *practical* scientists have proved that Darwin was right, and that evolution is *as certain* a law as that of gravitation. What a number of the 'learned' books of a few years ago opposing evolution must now be ignominiously withdrawn from circulation? And how *small* must the controversial parson and the lay evangelist, who would prove to you in 'two jiffies that science was all bosh,' feel at the thunders of competent scholars!'

While the Press is filled with suchlike articles, the people who do not think for themselves take for granted that science is right, and as a consequence, reject the Bible.

If I were asked to state the main cause of Modern Infidelity, I should say SCIENTIFIC FALSEHOODS INCULCATED AS TRUTH.

In the "Earth Review" for January, 1893, the following is found:

"HONEST AND NOBLE CONFESSIONS.

When we consider that the advocates of the earth's stationary and central position can account for, and explain the celestial phenomena as accurately, to their own thinking, as we can ours in Addition to which they have the evidence of their senses, and SCRIPTURE and FACTS in their favour. WHICH WE HAVE NOT: it is not without a show of reason that they maintain the superiority of their system...However perfect our theory may appear in our estimation, and however simply and satisfactorily the Newtonian hypothesis may seem to us to account for all the celestial phenomena, *yet we are here compelled to admit the astounding truth that,* IF OUR PREMISES BE DISPUTED AND OUR FACTS CHALLENGED, THE WHOLE RANGE OF ASTRONOMY DOES NOT CONTAIN THE PROOFS OF ITS OWN ACCURACY. — *Dr. Woodhouse, a late Professor of Astronomy at Cambridge."*

Those who believe the plain and provable facts of the Bible are set down as lunatics, but the above shows where the lunacy really lies.

John Wesley did not believe in the teachings of the men of the modern astronomical school, although most of his followers do. In his Journal he writes:

"The more I consider them, the more I doubt of all systems of astronomy...Even with regard to the distance of the sun from the earth, some affirm it to be only three, and others ninety millions of miles."

In Vol. 3 of the work which records his Journal, "Extracts from the works of Rev. J. Wesley," page 203, the following occurs:

January 1st, 1765.

"This week, I wrote an answer to a warm letter published in the 'London Magazine'; the author whereof is much displeased that I presume to doubt of the modern astronomy. I cannot help it; nay, the more I consider, the more my doubts increase; so that at present I doubt whether any man on earth knows either the distance or the magnitude, I will not say of a fixed star, but of Saturn or Jupiter — yea, of the Sun or Moon."

In Volume 13, page 359, referring again to the subject of theoretical astronomy, he says:

"And so the whole hypothesis of innumerable suns and worlds moving round them vanishes into thin air."

At page 430 of the same volume we find that:

"The planets revolutions we are acquainted with; but who to this day is able regularly to demonstrate either their magnitude or their distance, unless he will prove as is the usual way, *the magnitude from the distance, and the distance from the magnitude?*"

Thus, this admittedly great and good man stands out in bold contrast with many of the present day "reverend" gentlemen, The Bishop of Peterborough is another notable example. He says:

"I have no fear whatever, that the Bible will be found, in the long run, to contain more science than all the theories of philosophers put together."

Let me supplement this remark by stating that the Bible, and the Bible only, is THE scientific book of the Universe. It is the only volume which can be proved true from start to finish. I am not now going into the details of Bible Pyschology, Zoology, History, Philology, Ethnology, and the like. If time and space allowed all these could be proved as true as Bible Astronomy, and every one of them consistent with the facts of Nature, as I have shown Bible Cosmogony to be.

I shall now quote another infidel and reverend gentleman. In the *Christian World Pulpit,* of 29th March, 1893, the Rev. G. St. Clair, F.G.S., of Cardiff, contributes a sermon headed "Where is Heaven"; the text being taken from Acts i., 9: "And as they were looking He was taken UP, and a cloud received him out of their sight."

This wolf in shepherd's clothing goes on to say:

"In 1492 Columbus sailed westward to search of the East Indies, and 30 years later Magellan actually sailed away from Europe in one direction and returned in the other, having voyaged all round the world. It was thus shewn that the world is a globe. Previously the common notion had been that the earth was flat, and heaven a little way above the clouds, and the place of the dead — the wicked dead, if not all the dead — somewhere underneath. These were ancient ideas and the fact that we find them In the Bible is one proof that the Bible is an ancient book. The Bible writers had been educated to believe that God had laid foundations for the earth, or supported it on pillars. Heaven was His throne, the earth His footstool."

According to this preacher the Bible writers had been educated to believe a pack of lies. But, as I have already shown, what they believed, and what every consistent Christian believes to-day, is in perfect agreement with the great book of Nature, which lies open to every man who will believe its evidence.

Good advice is given to theologians by Dr. W. B. Carpenter in the "Echo" for 4th May, 1892, as follows;

> "If theologians will once bring themselves to look upon nature, or the material universe as the embodiment of the Divine Thought, and the scientific study of nature as the endeavour to discover and apprehend that thought, they will see that it is their duty, instead of holding themselves altogether aloof from the pursuit of science, or stopping short in the search for scientific truth, wherever it points towards a result that seems in discordance with their preformed conceptions, to supply themselves honestly to the study of it, as a revelation of mind and will of the Deity, which is certainly not less authoritative than that which He has made to us through inspired men, and which is fitted to afford its true interpretation."

Moses has been much maligned by modern scientific infidels. The "Muses" of December, 1895, has the following:

> "Moses has given his crude ideas as to the age of the world, but modern philosophers and scientists have clearly an equal right to give their deductions and opinions, especially as they produce evidence in which department Moses was very much at a disadvantage."

In the minds of unthinking multitudes science has carried all before it. as the following from Dr. Carpenter's work, "Nature and Man," pages 365 and 366, shows:

> "The geological interpretation of the history of the earth *has taken the place of the Mosaic Cosmogony* in the current belief of educated men, notwithstanding all the denunciations of theological orthodoxy."

The "Agnostic Journal," of 5th January, 1889, shows clearly that it is quite impossible to believe the Bible statements AND Modern Science:

> "The account of creation in Genesis is obviously inconsistent with the real facts, both as regards the relations of the earth to the sun, moon, and stars; the crystal vault separating the waters; the manner and order of succession of vegetable and animal life, and numerous other points. It can be defended only on the plea that *the inspired revelation was not intended to teach ordinary facts, such as those of astronomy and geology.*"

"The account of a universal deluge and the destruction of all life, except that of a few pairs of animals preserved and living together for a year in an ark of limited dimensions, from which the earth was re-peopled, involves not only physical impossibilities, *but is directly opposed to the most certain conclusions of genealogical and zoological science.*"

"The true history of the human race has been the direct contrary of that given by the Bible."

How long will it be ere professed friends of the Bible bestir themselves to read the book of Nature in order to discover whether the Book they profess to believe, because it gives evidence of its Divine Origin, is in accordance with the facts of Nature as we find them to-day r

The creed of the Agnostic—the know-nothing man—is briefly summed up by the "New York Independent" as follows:

"I believe in a chaotic Nebula self-existent Evolver of Heaven and Earth; and in the differentiation of this original homogeneous Mass. Its first-gotten Product which was self-formed into separate worlds, divided into land and water, self-organized into plants and animals, reproduced in like species, further developed into higher orders, and finally refined, rationalised, and perfected in Man. He descended from the Monkey, ascended to the Philosopher, and sitteth down in the rites and customs of Civilisation under the laws of a developing Sociology. From thence he shall come again, by the disintegration of the culminated Heterogeneousness, back into the original Homogeneousness of Chaos. I believe in the wholly impersonal Absolute, the wholly un-Catholic Church, the Disunion of the Saints, the Survival of the Fittest, the Persistence of Force, the dispersion of the Body, and in Death Everlasting."

Not only is there no room for God in what scientists, pleased to term "Nature," but there is no want of such Being, as the following from Carpenter's "Nature and Man," page 385, tells:

"'The laws of light and gravitation,' wrote Mr. Atkinson to Harriot Martineau 30 years ago, 'extend over the universe and explain whole classes of phenomena': this explanation, according to the same writer, is all sufficient, PHILOSOPHY FINDING NO GOD IN NATURE, NOR SEEING THE WANT OF ANY."

"The Earth and its Evidences," of 1st October, 1888, has the following:

"The attempt to harmonise the Mosaical and the modern or professional system of the universe, is plainly to attempt the communion of light with darkness. How often has failure waited on such incongruous unions! But, still, some there, are who seem to recognise the hopelessness of the task. They divest themselves of the idea that science must have been what justified in setting up her authority against that of the scripture records; — that humanity could not be so deceived as to adhere to a system of cosmogony, for more than a century and a half, which has been talked about and read and studied by some of the profoundest of modern thinkers, and to be proved, at last, no better than an old wives' fable, and as baseless and untrue, from the first line to the last, as if it had been invented by a class of village school children. If modern theories were only partially true, there might have been some consolation in thinking that humanity is doomed to err, and that the foundations or their vaunted science, were based upon facts. But this plea is utterly hopeless, and the very beginning of their complicated system i the most faulty of the whole. They are without excuse; for they deliberately abandoned the only clue given them at the ver. outset of their inquiry. The first chapter of Genesis supplied them with the outline of the entire system of physical cosmogony. That the earth was not a 'planet' was shown by the very first verse in the Bible. The two systems are kept most distinct throughout the whole of the sacred volume. The Almighty never calls himself the God of the sun or of the moon or of the stars; but in innumerable instances does he style himself the 'God of all the earth' the Lord and King of all the earth.' St. Paul declares that 'there are bodies celestial and bodies terrestrial, but the glory of the celestial is one, and the glory of the terrestrial is another.' This is so emphatically enforced through every page and chapter of the Bible, that to ignore or argue it away, is simply to treat the word of God as a lie from the beginning, to the end. If the universe is composed of nothing but planets, then the whole of a house is its roof, and the whole of the sea a dewdrop. All the planets were made on one and the same day, 96 hours after the creation of the earth. Many astronomers wonder why the earth was ever mentioned at all. 'A little insignificant dot of a planet,' about as proportionate in sine to the sun, as a honey-bee to a buffalo. And what is their authority for this astounding assertion — this impious contradiction to every word of inspiration? We ask what and who is their authority? Some Smith DC Jones or Robinson, that is all! And Christianity has bowed its head in meek submission to these upstart oracles, and treated the Word of God as dung, and with the same contempt that a philosopher would the intelligence of a magpie or a jay!

"Hugh Miller truly said that 'the battle of the evidences will have to be fought on the field of physical science and according to the logic of demonstrable facts,' This is the conflict to which I we are fast hastening, this the last great war of opinions, which every day is bringing nearer and nearer to our doors. The issues are most momentous, and as wide as the world in interest and importance. If 'science' wins the day, religion is the greatest bugbear that ever befooled humanity! If, on the other hand, the facts as narrated in the inspired records are infallibly and demonstrably true, then has Christendom been the victim of all the most impious and baseless imposture that ignorance and credulity could ever be exposed to.

"Modern science and religion cannot work together! Those who think they can cannot possibly believe or understand either! No man can eat bread and fancy he is drinking water. So no one can believe a single doctrine or dogma of modern astronomy, and accept the Scriptures as a divine revelation. And to teach them, side by side, in our schools and class rooms, is just to instill into the mind of the children that science is far superior to sense, and that falsehood and fraud are more desirable than truth and fact.

"Modern philosophy begins to attack the very first verse in the book of Genesis; and asserts that a pre-Adamite earth existed before the one subsequently referred to; that the seven 'evenings' and seven 'mornings' so accurately and particularly and distinctly specified in that first chapter, were not periods of twice twelve hours, but incalculable ages of time, of which no record exists, and are only made known to us through the laborious deductions of the more than inspired geologist! If this is so, then the 'seventh' day was an age also; and the Jews ought to have observed it, for a thousand years at a stretch! But if they were right in accepting it as a period of only 24 hours, then the remaining six must each have had exactly the same length, and the frantic geologist has to account for his 'stratas' and 'deltas' on some other supposition. It is important and highly necessary that we dwell a little on this, the first point that the modern theorist has assailed. If he can prove that he is right in his conjecture or rather in his positive assertion that days do not mean days, then is the infidel fully justified in laughing to scorn every other phrase and every other statement, from the first verse last in the Bible. And the theologian and the evangelist only expose themselves to derision and pity when they plead for any reverence for a book compiled on such vague and meaningless and delusive principles, and in language which has to be interpreted by pagan astrologers and infidel professors, before we can comprehend what is intended or ought to be understood! If the 'seven days' of Creation's week do not mean just what we understand by seven days, when all the Bible is symbolic, and is to be read upside down, and we must believe the very contrary to what is expressed.

"Till *after* the sixth day, all that was done, was not accomplished by any effort of nature, but by the personal agency of the Creator alone.

"Thus it is seen that Moses only begins to speak of Nature, or natural operations, *after* the seventh day. When, therefore, it is said that 'God rested,' it is, by natural implication, affirmed that Nature *began* to work or to act. And it is by losing sight of this most important fact that geology has made too many palpable blunders; and the soundness of that and all collateral sciences, in their very elementary principles, depend entirely on an accurate and distinct appreciation of this grand truth! The modern geologist may just as wisely argue that the live loaves that fed the five thousand, were made from grain that was ever grown in a field, or threshed in a barn, or ground in a mill, or baked in an oven, as to argue that what took place during those actual six days of Creation, was the effect of natural operations or of Nature's laws!

"Lord Bacon, in his 'Confession of Faith,' speaks most soundly upon this subject, as upon most others. He says, 'I believe, that God created the heaven and the earth; and gave unto them constant and perpetual laws, which we call "laws of Nature,' but which mean nothing but God's laws of Creation. That the laws of Nature which now remain, and govern inviolably till the end of the world, began to be in force when God rested from his work. That, notwithstanding that God both rested from creation since the first Sabbath, yet, nevertheless He doth accomplish and fulfil His divine will in all things, great and small, general and particular, as folly and exactly by providence, as He could do by miracle and new creation; though His working be not now immediate and direct, but by compass and control; not violating nature, which is He hath ordained for His creatures."

The inspired volume declares that:

"The works of the Lord are great, sought out of all them that have pleasure therein." — Psalm III, 2.

We are fully warranted, therefore, in seeking out the work of Nature, because, when rightly understood, God's works declare His wisdom and power. But the infidel works with the sole object of getting data for proving the work which so strongly testifies against his unrighteous myth and delusion.

In the Book of Genesis it is declared that God created the heaven and the earth, the lights in the heavens, the firmament to keep the waters above it from the waters below it, and in the books that follow, the foundations of the earth and other truths of like import are dealt with. The following passages show that the earth (dry land) is founded on the waters of the mighty deep, and is a motionless stretched-out structure, to which the heavens are parallel. Psalm 24: 1, 2; 136: 1-9; 102: 25; 104: 1-5; Isaiah 44: 24; 48: 13; 42:5; Deut. 5:8; Zech. 12; 1; Jeremiah 31: 35-37; 1 Sam. 2:81 Proverbs 3: 19; 8: 22-30; Job 9: 1-10; 38: 1-11.

The earth has borders which are impassable by man, as Job 26: 10 declares. See also Psalm 74: 16, 17.

The movement of the sun over a stationary world is clearly shown in such passages as Psalm 24; Ecc. 1: 5; Judges 5:31; Psalm 19.

That the stars are small is seen by the prophetic utterances of Revelations 6: 13. If they be worlds many times larger than the earth, how could they fall on it? See Rev. 8: 10.

Then 1 Corinthians 15: 40, 41, reminds us that there are terrestrial bodies as well as celestial, which truth the astronomer denies, by making the earth a celestial body:

"There are also celestial bodies and bodies terrestrial, but the glory of the celestial is one and the glory of the terrestrial is another. There is one glory of the sun, and another glory of the moon, and another glory of the stars; for one star differeth from another star in glory."

In Joshua 10: 12-14 the following language is utterly inconsistent with scientific teaching that the earth moves to cause day and night. If the sun stands still and Joshua commanded it to do what it always does, what an ignorant man he must have been, to be sure; To ask for a miracle to be performed in order that the "course of Nature" might remain as usual? Surely any person can see that it is totally unnecessary to ask the aid of miraculous power to prevent the sun from moving, if it never does move. But I shall let the passage speak for itself:

"Then spake Joshua to the Lord in the day when the Lord delivered up the Amorites before the children of Israel; and he said in the sight of all Israel, 'Sun, stand thou still upon Gibeon; and thou, Moon, in the Valley of Ajalon,' AND THE SUN STOOD STILL, AND THE MOON STAYED, until the people had avenged themselves upon their enemies...So the sun stood still in the midst of heaven, and hasted not to go down about a whole day."

Now, if the story of modern astronomy that the earth revolves and not the sun, be true, the only conclusion that can be arrived at is that the Bible is no better than a child's school book to record such an impossibility, and that, therefore, Joshua and the whole story is a myth. But we know that the sun moves, and we further know that the earth has neither axial nor orbital motion; and we conclude, therefore, that Joshua's command was perfectly consistent with fact and with his faith in the power of God to rule and overrule in His own world. Professor Totten, of Newhaven, in his pamphlet on "Joshua's Long Day," says:

"It is the Bible that Atheists and Infidels attack - the Old Testament chiefly - for they are logical, and perceive that if the foundation goes, the super-structure cannot stand, no matter how eloquently it can be clothed in Agnostic sermons...It will not do to doubt the universality of the Flood, and ask men to accept a Saviour who alludes to it...If the story of Eden and the Deluge, of Jericho and Joshua are myths or fables, and not literal facts, then to the still rational mind all that follows them is equally so, and faith, lost in those who foretold his Advent, can never be savingly and logically found again in Christ and his apostles."

These words are true, and show that modern astronomy and the Bible are on either side of an impassable gulf.

The Rev. W. Howard, of Liverpool, however, thinks differently. In his pamphlet "Joshua commanding the Sun to stand still; the miracle explained and defended," he says (*inter alia*):

"Why did not the ocean overflow the land? Run with a pail of water until you come in contact with a wall, and observe the effect upon the liquid, how it will dash over the side: and the sudden stoppage of the rotary motion of *the earth* would naturally send the sea almost all over the dry land...You know the shaking you get with the violent stoppage of an express train going at sixty miles an hour, and we ask you, please, to fancy the result to us, and to all cattle, dwelling houses, monuments, and even trees, if the *earth,* which at the equator *moves* nearly 1,100 miles an hour, was brought quickly to a standstill."

"I have now a FIFTH VIEW to lay before you, which appears to be both rational and simple."...My *belief* is this: Joshua and his men having walked all night, as the 9th verse tells us, would be tired next morning, but God caused a great trembling to spread itself amongst the foe, and there was an easy victory. When the war had pursued the Amorites some distance, hailstones fell upon them and did much damage. At the approach to Bethhoron the hailstorm increased in fury; and Joshua, seeing the devastation produced, and being cognisant of the fatigue of his men, *prayed Heaven to let the hurricane go on* till total and irreparable disaster was inflicted."

This poor man in his ignorance of the Bible and Nature tries to harmonise infidel astronomy with Bible truths, but he utterly fails, as the above quotation shows.

The learned Jewish historian, Josephus, in his "Antiquities of the Jews," Book v., cap. i, section 17, says:

"Joshua made haste with his whole army to assist them (the Gibeonites), and marching day and night, in the morning he fell upon the enemies as they were going up to the siege; and when he had discomfited them he followed them, and pursued them down to the descent of the hills. The place is called Bethhoron; where he also understood that God assisted them, which He declared by thunder and thunder-bolts, as also by the falling of hail larger than usual. Moreover, it happened that *the day was lengthened* that the night might not come on too soon, and be an obstruction to the zeal of the Hebrews in pursuing their enemies." ...Now that the day was lengthened at this time, and was longer than ordinary, is expressed in the books laid up in the Temple."

In a note under this paragraph, Mr. Whiston, the learned compiler of Josephus' works, while hesitating what explanation to give the miracle, says:

"The fact itself was mentioned in the Book of Jasher, now lost. Josh. 10: 13, and is confirmed by Isaiah (28: 21), Habakkuk (3: 11), and by the son of Sirach (Eccles. 46: 4). In the 18th Psalm of Solomon, ver. *ult.* it is also said of the luminaries, with relation no doubt to this and the other miraculous standing still and going back, in the days of Joshua and Hezekiah. 'They have not wandered from the day He created them, they have not forsaken their way, from ancient generations, unless it were when God enjoined them (so to do) by the command of his servants.' See Authent. Rec. part I., page 154."

The lights that God made for the use of this the only world, move above it, and in Joshua's long day the God of Creation hearkens to the voice of a man and causes the sun to stand still. *The miracle needs no defending.* IT ONLY NEEDS BELIEVING.

THE BIBLE IS LITERALLY TRUE (except in portions where it is very evident from the context that a symbolical meaning is to be attached to it) and MODERN ASTRONOMY IS ABSOLUTELY FALSE.

"Parallax," in his invaluable work "Zetetic Astronomy," says:

"To say that the Scriptures were not intended to teach science truthfully is, in substance, to declare that God himself has stated, and commissioned His prophets to teach, things which are utterly false. Those Newtonian philosophers who hold that the Sacred Volume is the word of God, are thus placed in a fearful dilemma. How can the two systems so directly opposite in character, be reconciled. Oil and water alone will not combine — mix them by violence as we may, they will again separate when allowed to rest. Call oil oil, and water water, and acknowledge them to be distinct in nature and value, but let no "hodge podge" be attempted, and passed off as a genuine compound of oil and water. Call Scripture the Word of God, the creator and Ruler of all things, and the Fountain of all Truth i and call the Newtonian or Copernican system of astronomy the word and work of man — of man, too, in his vainest mood — so vain and conceited as not to be content with the direct and simple teachings of his Maker, but must rise up in rebellion, and conjure into existence a fanciful complicated fabric, which being insisted upon as true, creates and necessitates the dark and horrible interrogative — is God a deceiver? Has He spoken direct and unequivocal falsehood? Can we no longer indulge in the beautiful and consoling thought that God's justice, love and truth, are uncharging and reliable as ever! Let Christians at least — for sceptics and atheists may be left out of the question to whatever division of the Church they belong, look to this matter calmly and earnestly. Let them determine to uproot the deception which has led them to think that they can altogether ignore the plainest astronomical teachings of Scripture, and yet endorse a system to which it is in every sense opposed.

"The following language is quoted as an instance of the manner in which the doctrine of the earth's rotundity and the plurality of worlds interfere with Scriptural teaching:

"'The theory of original sin is confuted (by our astronomical and geological knowledge); and I cannot permit the belief, when I know that our world is but a mere speck, a perishable atom in the vast space of creation, that God should select this little spot to descend upon and assume our form, and clothe Himself in our flesh, to become visible to human eyes, to the tiny beings of this comparatively insignificant world. Thus millions of distant worlds, with the beings allotted to them, were to be extirpated and destroyed in consequence of the original sin of Adam.

"'No sentiment of the human mind can surely be more derogatory to the divine attributes of the Creator, nor more repugnant to the known economy of the celestial bodies. For in the first place, who is to say among the infinity of worlds, whether Adam was the only creature tempted by Satan and fell, and by his fall involved all the other worlds in his guilt.'

"The difficulty experienced by the author of the above remarks is clearly one which can no longer exist when it is seen that the doctrine of a plurality of worlds is an impossibility. That it is an impossibility is shown by the fact that the sun, moon, and stars are very small bodies, and very near to the earth; this fact is proved by actual n on-theoretical measurement; this measurement is made on the principle of plane trigonometry; this principle of plane trigonometry is adopted because the earth is experimentally demonstrated to be a plane, and all the base lines employed in the triangulation are horizontal. By the same practical method of reasoning, all the difficulties which upon geological and astronomical grounds have been raised to the literal teaching of the Scriptures may be completely destroyed. The doctrine that the earth is a globe has been proved, by the most potent evidence which it is possible for the human mind to recognise — that of direct experiment and observation—to be *unconditionally false.* It is not a question of degree, of more or less truth, but of *absolute falsehood.* That of its diurnal and annual motion, and of its being one of an infinite number of revolving spheres, is equally false; and therefore the Scriptures, which negative these notions and teach expressly the reverse, must in their astronomical philosophy at least be *literally true.* In practical science, therefore, atheism and denial of Scriptural teaching and authority have no foundation. If human theories are cast aside, rejected as entirely worthless, and the facts of nature and legitimate reasoning alone relied on, it will be seen that religion and true science are not antagonistic, but are strictly parts of one and the same system of sacred philosophy.

"To the religious mind this matter is most important — it is indeed no less than a sacred question; for it renders complete the evidence that the Jewish and Christian Scriptures are absolutely true, and must have been communicated to mankind by an anterior and supernal Being.

"If, after so many ages of mental struggling, of speculation and trial, of change and counter-change, we have at length discovered that all astronomical theories are false; that the earth is a plane and motionless, and that the various luminaries above it are lights only and not worlds; and that these very facts have been declared and recorded in a work which has been handed down to us from the earliest times — from a time in fact, when mankind had lived so short a period upon the earth that they could not have had sufficient experience to enable them to criticise and doubt, much less to invent and speculate — it follows that whoever dictated and caused such doctrines to be recorded and preserved to all generations must have been superhuman, omniscient, and to the earth and its inhabitants pre-existent. That Being could only be the Creator of the world, and His truth is recorded in the Sacred Writings. The Scriptures — the Bible, therefore — cannot be other than the word and teaching of God. Let it once be seen that such a conclusion is a logical necessity; that the sum of the purely practical evidence which has been collected compels us to acknowledge this, and we find ourselves in possession of a solid and certain foundation for all our future investigations.

That everything which the Scriptures teach respecting the material world is *literally true* will readily be seen. It is a very popular notion among astronomers that the stellar universe is an endless congeries of systems, of suns and attendant worlds, peopled with sentient beings analogous in the purpose and destiny of their existence to the inhabitants of this earth.

"This doctrine of a plurality of worlds, although it may be admitted to convey most magnificent ideas of the universe, is purely fanciful, and may be compared to some of the 'dreams of the alchemists' who laboured with unheard-of patience and enthusiasm to discover a 'philosopher's stone' to change all common metal into gold and silver; an elixir vitae to prevent and to cure all the disorders of the human frame; and the 'universal solvent' which was deemed necessary to enable them to make all things homogeneous, as preliminary to precipitation, or concretion, into any form desired by the operator. However grand the first two projects might have been in their realisation, it is known that they were never developed in a useful and practical sense. They depended upon the third — the discovery of a solvent which would dissolve everything. The idea was suddenly and most unexpectedly destroyed by a few remarks of a simple but critical observer, who demanded to know what service a substance could be to them which would dissolve all things. Seeing that it would dissolve everything what would they keep it in! It would dissolve every vessel wherein they sought to preserve it. The alchemists had never 'given a thought' to such a thing. They were entirely absorbed with the supposed magnitude and grandeur of their purposes. The idea never struck them that their objects involved inconsistency and impossibility; but when it did strike, the blow was so heavy that the whole fraternity of alchemists reeled almost to destruction, and alchemy as a science, rapidly expired. The idea of a 'plurality of worlds' is as grand and romantic as that of the 'universal solvent' and is a natural and reasonable conclusion drawn from the doctrine of the earth's rotundity. It never occurred to the advocates of sphericity and infinity of systems that there was one great and overwhelming necessity at the root of their speculations. The idea never struck them that the convexity of the surface of the earth's standing water required demonstration. The explanation its assumption enabled them to give of natural phenomena was deemed sufficient. At length, however, another 'critical observer' — one almost born with doubts and criticisms in his heart — determined to examine practically, experimentally, this fundamental necessity.

"The great and theory-destroying fact was quickly discovered that the surface of standing water was perfectly horizontal. Here was another death-blow to the unnatural ideas and speculations of pseudo-philosophers.

"Just as the 'universal solvent' could not be preserved or manipulated, and therefore the whole system of alchemy died away, so the necessary proof of convexity in the waters of the earth *could not be found,* and therefore the doctrine of rotundity and of the plurality of worlds must also die. Its death is now merely a question of time."

Appendix

—

The Earth an Irregular Plane

——

By William Thomas Wiseman. F.R.G.S., &c.

——

The surface of all water, when not agitated by natural causes, such as winds, tides, earthquakes, &c., is perfectly level. The sense of sight *proves* this to every unprejudiced and reasonable mind. Can any so-called scientist, who teaches that the earth is a whirling globe, take a heap of liquid water, whirl it round, and so make rotundity? He cannot. Therefore it is utterly impossible to prove that an ocean is a whirling rotund section of a globular earth, rushing through "space" at the lying-given-rate of false philosophers.

When a youth, I stood upon the Dover shore of the English Channel, and was told to watch a departing ship. "See! There she goes; down, down, down! The hull has disappeared! She is out of sight! Now, my boy, you have had an occular demonstration that the world is round (meaning globular in shape), and SEEING IS BELIEVING." I walked up to an "old salt" who had a telescope, and said: "Can you see that big ship through your glass that's gone down the Channel, and is now out of sight?" "Yes, my son. Look!" The big ship immediately came into view again, as I peered through the sailor's glass! "Why! my — told me the earth was round, because that ship I can now see had turned down over the horizon!" "Aha! aha! sonny, I know they all says it! Now, I have been all over the world, but I never believed it. But, then, I have no learning, only my senses to rely upon, and I says SEEING IS BELIEVING."

I now, after many years, endorse the old sailor's experience, that the world is not a globe, and I have never found the man who could prove by any practical demonstration that he, or I, are living on a whirling ball of earth and water! Nor does the *dense* earth and the *rare* air rush round together! Declare, ye scientists, IF YOU KNOW! The Scriptures of God's inspired Prophets contradict the unreasonable illogical, unscientific delusion, and false philosophy, that the fixed earth is a hollow fireball with several motions!

> "There is an old adage, by which you can fix them,
> There is not one lie true, no, not if you pick them."

Evolution

When grovelling minds of little worth
Forsake the Lord of heaven and earth,
What dreams of fancy they imbibe;
They claim as kin the monkey tribe.
They set all history at defiance
And call their speculations science,
Then try to shew the wondrous plan
Of how the ape became a man.

All things to God men used to trace,
And every species kept its place.
But now we're told that men and worms
Have only sprung from lower forms;
And when proud science lends her aid
They'll tell us how these forms were made;
This thought is theirs—O happy notion!
"Mind is but matter put in motion."

In works of art they see design.
And own that wisdom did combine;
They say you may behold it in
A watch, a mouse-trap, or a pin;
But all the flowers that scent the breeze,
The fruits that grow upon the trees.
The wondrous form and powers of man,
Arose, they say, *without* a plan.

If science shews that man escapes
And leaves the ranks of grizzly apes;
Then science may reverse the plan
And prove the ape a fallen man.
And this new species yet may boast
And gain the tails their fathers lost;
As matter moves and beauty withers,
Time yet may class them with their fathers.

No God they see in all creation;
They spurn the thought with indignation,
Their main pursuit in life' is pelf;

Their creed is — *"Always mind yourself."*
They say to saint and sage and ruffian —
"The future state is but a coffin;
And when we pass beyond life's storms.
We hope to be devoured by worms.

O charming hope for which they wait!
What glory gilds their future state!
If here they do but little good,
Yet after death they're used as food.
Then let this glowing prospect cheer.
Take care of self while you are here,
Grow fat and plump till latest breath.
And you'll be useful after death.

D. S.

From the "Christian Commonwealth" Jan. 25th, 1894.

The New Scriptures

According To Tyndall, Huxley, Spencer and Darwin.

1. Primarily the Unknowable moved upon comos and evolved protoplasm.

2. And protoplasm was inorganic and undifferentiated, containing all things in potential energy; and a spirit of evolution moved upon the fluid mass.

3. And the Unknowable said, "Let atoms attract" and their contact begat light, heat, and electricity.

4. And the Unconditioned differentiated the atoms, each after its kind; and their combinations begat rock, air, and water.

5. And there went out a spirit of evolution from the Unconditioned, and working in protoplasm by accretion and absorption, produced the organic cell.

6. And cell, by nutrition, evolved primordial germ, and germ developed protogene; and protogene began eozoon and eozoon begat monad, and monad begat animalcule.

7. And animalcule begat ephemera; then began creeping things to multiply on the face of the earth.

8. And earthly atoms in vegetable protoplasm begat the molecule, and thence came all grass and every herb in the earth.

9. And animalcule in the water *evolved* fins, tails, claws and scales; and in the air, wings and beaks; and on the land they sprouted such organs as were necessary, as played upon by the environment.

10. And by accretion and absorption came the radia and mollusca, and mollusca begat articulata, and articulata begat vertebrata.

11. Now these are the generations of the higher vertebrata, in the cosmic period when the Unknowable evoluted the bipedal mammalia.

12. And every man of the earth, while he was yet a monkey, and the horse while he was a hipparion, and the hipparion before he was an oredon.

13. Out of the ascidian came the amphibian and begat the pentadactyle; and the pentadaclyle, by inheritance and selection, produced the hylobate, from which the simiadae in all their tribes.

14. And out of the simiadae the lemur prevailed above his fellows, and produced the platyrhine monkey.

15. And the platyrhine begat the catterhine, and the catterhine monkey begat the anthropoid ape, and the ape begat the longimanous orang, and the orang begat the chimpanzee, and the chimpanzee evoluted the *what is-it?*

16. And the what-is-it went to the land of Nod, and took him a wife of the longimanous gibbons.

17. And in process of the cosmic period were born unto them and their children, the anthropomorphic primordial types.

18. The homunsulus, the prognathus, the troglodyte, the autochthon, the tarragen, these are the generations of primeval man.

19. And primeval man was naked and not ashamed, but lived in quadrumanus innocence, and struggled mightily to harmonise with the environment.

20. And by inheritance and natural selections did he progress from the stable and homogeneous to the complex and heterogeneous; for the weakest died and the strongest grew and multiplied.

21. And man grew a thumb, for that he had need of it, and developed capacities for prey.

22. For, behold the swiftest men caught the most animals, and the swiftest animals got away from the most men; wherefore the slow animals were eaten and the slow men starved to death.

23. And as types were differentiated the weaker types continually disappeared.

24. And the earth was filled with violence; for man strove with man, and tribe with tribe, whereby they killed off the weak and foolish, and secured the survival of the fittest.

From the "Rainbow" and copied from, an American Journal.

Truth Will Conquer

Dedicated to the Members of the Church Congress, held at Norwich, 1895.

"Ah, man!
You are so great — too great for this small world,
For you have proved that Christ is all a lie!
The Gospel that He taught us but a 'MYTH,'
The Bible but a pack of legends, old
And false traditions — you can prove it. Ay,
You are so wise. O vain, presumptuous man,
You love to think the 'Word of God' is false,
And hope to mar its beauty with your sneers.
Rail on; God's citadel shall never fall to you,
Smite as you may.
Ah, 'Science,' SOURCE OF INFIDELITY,
You blazon great discoveries to the world,
Fresh wonders brought to light by such as you,
Revealing Nature's laws' (*we* call them God's),
Proving all things exist by hidden sacred laws,
And, adding pride to folly, call them *'chance.'*
Fool! God has made those laws, and set the sun
And all the planets daily to perform
Their wondrous course, through endless aeons on,
From cycle unto cycle, ne'er to cease,
Do ye not know that what has been *shall be,*
That nought is new, nought underneath the sun.
As said the King of Wisdom — Solomon?
But, ye, the more ye search, new wonders find.
And newer wonders, till the less ye love
The Wonder-Maker, All Creating God.
Why is it thus? and why does Wisdom (?) turn
Your heart from God, when He all Wisdom is?
But ye will rave in your demented pride,
Wise in the worldly wisdom of the world,
Wise in your darling theories — so false
To sense, or truth, or manly, honest doubt
Ye know so much, and yet one little child,
In her sweet faith, is wiser than ye all,
And nearer unto God. And ye would force
Your base *opinions* on the ears of men,

And bid them hearken to your hollow words!
Leading the blind with your phantasmal talk,
Yourselves more blind than they, more dull your sense;
False prophets, fools, to kick against the pricks
As did the bigot Pharisees of old!
But ye may rave; think ye that truth will fail?
Think ye with puny breath to blast the Rock
That has stood firm for nineteen hundred years
Against the sceptic's scorn, the mocker's laugh,
And borne the brunt of Infidelic sneer
Immutable, in majesty supreme?
Watching you beat yourselves to death upon it!
We fear not: do your worst. *Right* conquers *Might.*
And God's great *Truth must* conquer in the end!"

John Merrin.

The Glory of God

The inspired Psalmist says that "The heavens declare glory of God; and the firmament showeth his handiwork"; therefore, whatever some professed Christians affirm to the contrary, the subject of Creation is connected with right views of God. His worship, and His glory. But if we would have a right conception of God, and His glory, we must see to it that we have a right conception of His works Creation, How, for instance, do we obtain an insight into the character of any great man, whether he be a poet, Politician, sculptor, general or king? Is it not by his acts, or his works? But suppose these acts, or works, are misrepresented to us, or defaced by someone, should we not have false and distorted views respecting the author, artist, or the maker of those things? Assuredly. And so it comes to pass in respect to the construction of the world, false views of the universe have led men into a misconception respecting the character of God, and even alas! in many cases to a denial of the very existence of such a personal Being.

Let us, then, endeavour to come back to first principles. The world exists, and must have come from somewhere. It" is "unthinkable" to say it came by chance, or any "fortuitous concourse of atoms," Its wonderful variety, the general co-relation and adaptability of its various parts, and the exact and never-failing motions of all the heavenly bodies, *prove,* to any well-balanced and unprejudiced mind, that some grand and controlling intelligence directs and rules over all. As the apostle Paul declares, "The invisible things of Him from the creation of the world are clearly seen, being understood by the things that are made, even His eternal power and Godhead; so that they are without excuse." Rom. i., 20.

A grand truth ties in this statement of the apostle. Paul was no fool. It is allowed on all sides, alike by friend and foe. Sceptic and Christian, M. Renan and the Archbishop of Canterbury, that no one man has had more influence in forming Christianity, the history of which has for eighteen centuries been making the history of the civilised world, than the apostle Paul. His name will be had in honour when the names of the adversaries of the truth will have sunk into merited and everlasting oblivion. And this great man agrees with the Psalmist in leaching that the Creation, as set forth in the Bible, and as found in what some call "Nature," sets forth unmistakably the grand truth that God IS. Now, this is a fundamental verity, and the foundation of all true faith. GOD IS, And "he that cometh to God must believe He IS, and that He is a rewarder of them that diligently seek Him." Now, this faith is, on the one hand, neither an unreasoning credulity, nor, on the other hand, is it a bigoted *dis*belief. It is based on an intelligent and reasonable understanding of the things that are seen above and around us.

The Book of Nature is open to all men; but it must be read and studied without prejudice and without philosophical bias. We must come to it like little children, with the honest desire to know the truth, and not attempt to read into it our own, nor anyone else's, plausible or implausible hypotheses. If we do this patiently and persistently, we shall be "rewarded": the grand and ineffaceable truth will dawn upon us that GOD IS.

We shall see His glory in the bright and blazing sun as he goes forth majestically, like a giant to run his daily course, We shall own *His* Power and Godhead when the moon, queen of the night, rises in quiet and stately splendour, to reflect her silver radiance in every rippling stream. And we shall confess *His* wisdom and unfailing skill when, at night, we gaze up into the firmament and behold ten thousand glittering gems, shining in matchless beauty, and shedding upon the earth their silent influences, as they nightly perform their appointed revolutions. Truly we shall then confess with the Psalmist, that "the heavens declare the glory of God, and the firmament sheweth His handiwork."

"The firmament sheweth His handiwork." That vast and incomparable structure which spans the heavens, and covers the earth with its capacious dome, divides the waters which are "above" the firmament from the waters which are "under" the firmament. And when we realize something of the tremendous size of this tent-like covering, spanning with one mighty arch across the whole of the outstretched earth; when we consider its weight, its strength, its stability, and the avowed purpose for which it was made by the Creator, we can unhesitatingly and devoutly again exclaim with the Psalmist,

"The firmament sheweth His handiwork." No wonder such a "work" occupied the whole of one day, the third, in the "great and marvellous" work of the six days Creation. Job, one of the finest, and certainly one of the most ancient, of true philosophers, when comparing the works of God with the puny works of man, asks: "Hast thou with Him spread out the sky, which is strong, and as a molten looking-glass?" Job 37: 18. It is, perhaps, this mirror-like quality which the firmament possesses that makes unbelieving "scientists" think that they can, with their glasses, peer into what they call "space," which they affirm to be "boundless." As well might a child, gazing upon the bosom of a glassy lake, affirm that it had no bottom, and that the sky and clouds, reflected from its placid surface, were slumbering in the unfathomed depths below, and not above, its waters.

The idea of illimitable "space," filled with an infinity of revolving worlds or globes, is not only a bewildering idea, unfounded on fact, but it directly tends to remove the Creator, or rather the idea of a Creator, far, and farther, away from this earthly plane of ours. It necessarily and logically leads to Atheism; and too often, alas! it practically leads men there. The idea of Heaven as a place, the abode of The Eternal, becomes to the logical and thinking Newtonian a *myth;* and God, if he acknowledge such a personal Being at all, becomes farther and farther removed from the scene of all earthly operations. Whereas the Saviour of the World, who "came down from Heaven," to do his Father's will, taught His disciples to believe that Heaven was not very far off; that it was directly and always "above" us; that God was concerned in the work of His hands; and that as "our Father," He was near enough to hear the prayers of all those who call upon him in sincerity and truth. This is assuring: this is comforting. God cares for the world; and He will punish those who afflict mankind with their selfishness, their greed, their falsehoods, and their oppressions. Yea, God has "so loved the world" — not the "globe," as some misguided Christians have lately printed and perverted this sublime text with a ridiculous "globe" stamped on the paper — God "so loved the world that He gave His only begotten Son, that whosoever believeth ill Him should not perish but have everlasting life." This, we say, is comforting. It is assuring. But, on the astronomical hypothesis, the world is like an uncared-for orphan, or a desolate wanderer: God is removed too far from us to be of any practical use; and the idea of Heaven is so vague, that such a place, if it exist at all, may be anywhere or nowhere; "all round the globe;" or spirited away from us altogether, "beyond the bounds of time and space." Thus the Christian's hope is undermined, and his faith is eaten away at the very core by this insidious and so-called "scientific" worm. This is most calamitous; yet even some of our "spiritual guides" are either so false to their professions or are so deceived themselves, that they cry out: "It does not matter what shape the earth is; we don't care whether it be round, or flat, square or oblong, so long as" — yes, so long as they get a good "living," and hold a respectable position in society? Is this it? Such a confession really means, when put into plain language: We do not care whether the Bible be false, in its record of Creation, so long as our interests or our hope of "Salvation" is assured. But "woe" is pronounced against such easy going shepherds of Israel. "Woe" to them who are leaving their flocks to become a prey to the devouring wolves of "Science," "falsely so called," as the great apostle intimates. Let us be on our guard. There are honourable exceptions to such false shepherds and teachers, and others are being raised up to warn us. We have quoted some of their noble testimonies. Let us, give heed to these needful warnings. God has never left Himself without witnesses to this Truth whether in Nature or in Revelation. We may shew this, if the Lord permit, more fully another time as regards Creation truth.

In conclusion, we would call the attention of all our readers to the seasonable warning given us by the Apostle Paul, where he says, "Beware lest any man spoil you through philosophy und vain deceit, after the tradition of men, after the rudiments of the world, and *not after* CHRIST." Col. 2: 8. And again, Let us prove all things; and hold to that which is good."

"Zetetes."

How Old Is the Earth

By Alex McInnes

A squabble over the earth's age lately broke out between Lord Kelvin, styled by Earl Salisbury, "the greatest of living scientists," and a Professor Perry, who disputed the infallibility of his chief. The scientific lord, formerly William Thomson, assumed, or as usual supposed, that the earth is a "homogeneous body," cooling at a fixed and uniform rate| therefore, that its age is somewhere between 20 millions and 400 million years. However, the lordly dictator having established his supposition, larded over with mystical mathematics, also in words of thundering sound, what multitude of simpletons will now gulp down the bolus without ever asking for the evidence so wholly wanting. Now, is a university professor so blind as not to see the enormous difference between 20 millions and 400 millions - viz., 380 millions, to count which at the rate of 60 per minute, 12 hours daily, would occupy 24 years of a man's life? Then, why call the vast continents making- up the land or earth a body, seeing that they have neither head, legs, nor any such members; and why a body any more than a soul? But, if by earth is meant all the oceans and continents rolled together into an astronomer's imaginary globe, land being solid and ocean fluid, where is the homogeneity? *En passant,* this misuse of the words body and earth are but specimens of the wholesale verbal jugglery practised by scientists to cause mental confusion and darkness. Moreover the Glasgow professor to make the earth's age what he pleases has only to assume the rate of cooling accordingly. Yet the 400 million years being too paltry a period for the evolution fable. Professor Perry rejects the supposition of cooling, and assumes that the earth's centre is now in a highly molten state, and with as much confidence as if he had been down in the infernal regions making a personal inspection, whilst Lord Kelvin assumes a familiarity with the earth's primeval conditions as if he had witnessed the Creation.

Is not the fabulous chronology after all like the ocean-land-globe, a mere heirloom of ancient heathendom: The Japanese and Chinese to make chronology square with their abominable Buddhism suppose 3 million years for the earth's duration, the Hindoos for Brahminism 6 millions; and now Professor Thomson, to please the atheistic evolutionists, is even willing to grant 4,000 million years as the greater limit, thereby confessing a blunder of 3,600 millions!

Further, the scientists can see nothing to admire beyond or above what they call nature, that is, the visible Creation, which by *their assumption* is its own Creator — having had an eternity of ages to evolve sun, moon, stars, oceans, and continents out of an *imaginary* fiery gas — a god unaccounted for; life out of death; order, beauty, light out of darkness and chaos; many thousand kinds of plants out of granite; thousands of kinds of beasts, birds, fishes, insects, out of cabbages, trees, &c., and man out of no one knows which kind of monkey! Still this goddess Nature is confessed to be as helpless as the puppet of a Punch and Judy show, being entirely dependent on mythical laws which act with an energy too omnipotent for Nature to resist, and she is pulled, hurled, tossed, evolved, exploded, just as these mythical laws please. Again, the laws themselves are under a necessity of operating according to rules, fixed how, why or when, no one knows; yea, unchangeable, at least, since tadpoles grew out of cabbages to father our ancestral apes, gorillas, or baboons. *But whence the* INVOLUTION *that must have* PRECEDED *the* EVOLUTION is another nut too hard for scientists to crack!

Is it hard with such cunning fables to deceive the multitudes so debased by the lying stories and abominable idle gossip of newspapers and like literature? And though foolish editors may jest at Moses, yet the Pentateuch still stands the oldest historical monument, so well authenticated and so full of unassailable internal evidence — so plainly endorsed by Jesus, whose well-attested Christhood no lover of truth can deny. With the date of Creation given in Genesis, as well as the Patriarch's ages, along with periods of time given by the sacred Hebrew historians following Moses, we may calculate down to the first year of Cyrus, where we are assisted by Josephus and Greek historians, thereafter by an unbroken chain of literature down to the present year, eclipse and transit cycles confirming all. Hence we know that about 6,000 years ago, God said, "Let there be," and there was.

In Dr. Dick's "Natural History" we have a specimen of the Geological method of calculating. He *supposes,* of course without any proof whatever, that God did not make the bed of the Niagara, but that that river cut for itself the passage of six miles below the falls; and further *supposing* the Niagara to cut one foot yearly, he concludes it must have been so working for 31,000 years, but if it cuts, as others *suppose,* one inch yearly, we have more than 300,000 years as the present or quartary period. Next he *supposes,* still without proof, that the underlying systems, the tertiary, secondary, primary, primordial rocks, represent as many antecedent periods of time. So, the quartary being 500 feet thick, and the tertiary 3,000 feet, we have six times 31,000 years or six times 300,000 years to add for the earth's duration. Again, the thickness of the secondary rocks being 15,000 feet their period must be 30 times that of the present; whilst the thickness of the primary is three times, and that of the primordial five times that of the secondary. Therefore, the earth's age is somewhere between 8½ millions and about 100 million years; without taking into account the unknown period of the igneous rocks. However, we now from Genesis 1 that God made all things in six days, all the rocks on the third day, in strata according to Job xxxviii. 5; therefore, granting the Niagara to cut one inch early it must since the creation have worn away only 600 inches or 50 feet.

Accordingly, shall we compute the earth's age by the age and contradictory guesses of fellow worms called geologists, or by the authority of the Creator Himself?

Evolution—What Does It Mean?

One school in attempting to bridge o'er the chasm,
Invented the germinal cell "Protoplasm,"
Which was first organic, but afterwards seen
To grow into "Sponges" and "Polyps" marine;
From thence by "Absorption," "Accretion," and growth,
Giving birth to the "Bivalves" or "Molluscs" or both.
These creatures by striving grew fins, tails and claws,
In spite of Dame Nature's implacable laws.
They sprouted and turned into reptiles amphibious;
Of obstacles placed in the way quite oblivious.
Urged on by "Necessity" upwards they grew,
Day by day giving birth to some quadruped new,

Evolving, re-forming without intermission
"As played upon by the surrounding condition."
Then "Like produced *un* like" without hesitation,
Earthy atom transformed into rich vegetation.
Animalculae left their aquatic abode,
And into the Forests by thousands they strode.
Frogs changed into birds at the voice of the Sirens,
And everything living "changed with their environs."
The Lichens from every restriction then broke,
And evolved both the Lepidodendron [1] and Oak.
'Twas a wonderful time and a wonderful sight
To see how each day brought new objects to light.
The stratified rock the strange story relates,
How the "Invertebrata" [1] begat Vertebrates;
And the "Ichthyosaurus" [1] one night in a freak,
Gave birth to the "Mastodon" [1] — (minus the beak),
While the tidy Acidian evolved from the Oyster,
Emerging somewhat like a monk from his cloister
The Bear from the Mole in the past we descry,
While the Bumble Bee came "by descent" from the Fly.
Then the Lemur begat the grim Ape Catarrhine,
From thence came the others "in process of time."
Their tails being "chaffed," became shortened, 'till soon
We arrive at the hairy-faced, tail-less Baboon.
These quarrelled and fought in the Forests primeval,
Impelled by an inherent spirit of evil.
The Pentadactilians ignoring all trammels,
Produced the most curious Terrestrial Mammals;
While the Porpoise and Sea-Horse plunged into the deep,
Determined henceforward to water to keep.
"By the use and disuse" of their parts, as it suited,
They wandered (to no spot particular rooted),
One half the world took with the other to strive,
'Till naught but the "Fittest" were found to "Survive."
At last Man appeared; but, amazingly strange!
From that moment the animals never could change.
"Like" at last "produced like," and the laws became fixed,
Which explains why the Species since never got mixed.

J. W. H. *From "The Anti-Infidel," March, 1887*

[1] These are fossil animals and plants.

Our Earth Motionless

—

Definite Conclusions of Science.

—

A Popular Lecture proving that our Earth neither rotates upon its axis nor around the Sun. — Delivered at Berlin by Dr. Shoepfer.

—

Gentlemen, — One should be endowed with unlimited courage to dare come out before a large audience with proofs of the erroneousness of a scientific formula which since our earliest youth we had been taught to regard as the only correct and unerring theory. I am pretty certain that at this moment you have come to the same conclusion about me, as four months ago, I would have entertained myself of any man who should have asserted that it is not the earth which revolves around the sun, but the sun which revolves around the earth. I would have considered such a man either an ignoramus or a lunatic; nevertheless. I now consider the immobility of the earth an incontrovertible fact, and even hope that my convictions will be shared by those who without prejudice will reflect upon that which I will now impart to them.

Some time ago we had the opportunity of witnessing the series of experiments with a pendulum which, according to the theory of the celebrated physicist, Leon Foucault, furnish proof of the diurnal rotation of the earth around its axis, I had long neglected to acquaint myself with these experiments, although, while explaining to my pupils the motion of the earth around the sun, I had always found very extraordinary results — absurd, I ought to say — one circumstance pertaining to this motion with which you will acquaint yourselves in my present lecture. So firm was my conviction of the diurnal and annual revolutions of our globe (earth?) that I had accepted even Foucault's experiments with the pendulum as sufficiently demonstrative.

Meanwhile, I had been appointed to assist in the experiments, and, as they bear directly upon the subject in hand, I will briefly state in substance the results.

If, choosing any given point in space near our globe, we imagine a limitless series of circles, then, in consequence of their parallel position to the equator, we term such series of circles parallels.

From the exterior form of the earth we conclude that these circles go on diminishing as they near the poles. If we fancy two such circumterraneous parallels as dividing this auditorium, then the northern parallel will be shorter than the southern. In the rotation of the earth around its axis in 24 hours both parallels will have to accomplish the rotation in the same space of time; and as they complete the circuit simultaneously, but the southern parallel is longer than the northern, then, consequently, every point of the southern parallel must move with greater velocity than the like points of the northern.

Let us now throw a glance on the apparatus called the pendulum, which is well-known to every one, but in the particular case in point a very equivocal authority. It is easy to demonstrate that the arc of the vibration of the pendulum does not depend upon the change (Drehung) of the point of suspension. This undisturbed regularity of the vibration of the pendulum has served M. Leon Foucault as a proof of the rotation of the earth around its axis. If we cause such a pendulum to vibrate across the parallels which we are imagining to pass through our audience, then the arc of the vibration, as Foucault tells us will (not) change from the axial rotation of the emplacement, and will begin, in consequence of this, to gain in rapidity on the northern and less rapidly moving parallel, and will be outstripped by the southern one, which moves quicker. In such a case, the arc of the pendulum will soon diverge from its direction from north to south, and its point turned to the north will near the east, and with the point turned south will begin more and more to near the west, till, finally, the pendulum will change its motion in the direction from east to west.

Now the reason for a deviation of the pendulum has ceased; it vibrates no more across two parallels, but only across one. The cause of its deviation from its first direction is removed; it would then seem that the deviation itself ought not to take place any longer, but nevertheless it still continues. The pendulum abandons the east and west direction to approach with its points the southeast and northwest until it reaches its starting point, at which it must again deviate according to Foucault's theory.

As the pendulum does not preserve the direction from east to west, but always gets farther and farther away, I conclude that the deviation of the pendulum is not caused by the axial motion of the earth, but is due to some other motion yet unknown.

By a series of careful experiments I have found that all pendulums are not liable to a deviation in the same degree; the heavier the ball, the more rapidly it will deviate. And as the rotation of the earth around its axis — if we admit its existence — ought to be manifested everywhere equally, then its deviation also, for every kind of pendulum, must be equal in time; but this in reality is just what is not the case.

The conviction that Foucault's arguments were erroneous forced me to verify at the same time all other proofs which have hitherto been regarded as demonstrating the rotation of the earth around its axis, and it was then I found that we had no evidence for such a theory.

Already in antiquity Aristarchus of Samos and other philosophers, several centuries before Christ, affirmed that the stellar sphere is motionless, and that the daily rising and setting- of the stars can only be accounted for on the theory of the earth's rotation around its axis. But all these men, profound thinkers, had come to the above conclusion only from the fact that otherwise such an incredible rapidity of the celestial bodies as would enable them to accomplish a diurnal circuit around the earth could never be accounted for. Of course every one must agree with me that at the present moment such an argument would be regarded as very small proof. Indeed, if we were able to take a little peasant boy from a country in which railroads were unknown and tell him of the existence of carriages which are able to make a mile in five minutes, of course he could never believe us; such rapidity would seem incredible to him. He is ignorant that light travels with a velocity of 40,000 miles a second, and that the rapidity of electricity is still more considerable! Thus, this argument with respect to the celestial bodies whose nature is as yet so little understood, and the path of whose motion is a vacuum or in a space filled with attenuated matter is only assumed or guessed at upon the strength of an hypothesis — that these bodies cannot have such a velocity of motion as to be able in twenty-four hours to circumscribe the earth — such an argument, to make us reject the possibility of the rotation of the celestial sphere, is certainly weak and futile.

But the contrary position, the one commonly accepted, also proves untenable when we look into it carefully.

It was found in the measurement of the earthly meridians that the globe is flattened towards the poles, and that in consequence of this, the equatorial diameter is greater than the line which passes through the axis of the earth from one pole to the other. Man, who endeavours to penetrate into all the mysteries of nature, tried to find the reason for such a flatness, and then comes Newton and explains it by the rotatory motion of the globe. In consequence of such a rotation all the component parts of the earth, and especially the bodies to be found upon its surface, receive an impulse to abandon the earth. Such an impulse is then named the centrifugal force.

At the poles, where the rapidity of motion is equal to 0, that force is also equal to 0; further from the poles to the equator that force increases in ratio with the increase of the parallels, so that the greater the parallel is, the more rapidly as I have already said, must move each of its points. In consequence of this, they say, the greater part of the earth's mass is gravitating toward the equator; and for the same reason, the centripetal force, acting on the equator with greater intensity, compels the concentration there of the greater portion of the mass. Hence it is finally concluded that the earth must forcibly rotate around its axis, because were there no such rotation there would be no centrifugal force, and without such a force there would exist no gravitation toward the equatorial diameter or zone.

We have laid before you now one of the existing evidences of the rotation of the earth, I do not accept such an argument, but reject it with many other scientists who have discarded it before myself...

Therefore, gentlemen, until we have more weighty argument to explain satisfactorily the accumulation of the mass of the earthy matter on the warmer zones, I cannot undertake to accept as a reason for it a certain centrifugal force, appearing as a consequence of the motion of the earth around its axis, and I will not allow the hypothesis, were it but because I know beforehand to what inexplicable contradictions this centrifugal force would bring us. Some of these I will point out presently.

We must now consider the fourth and last evidence of the rotary movement of the terrestrial globe.

In 1867, M. Richer remarked that a clock of his, which kept good time in Paris, having been transferred to Cayenne, *i.e.,* five degrees north of the equator, began to lose two and a half minutes daily. Richer had to shorten the rod of the pendulum one and a quarter lines to make the clock go right. It is well-known that the time of the vibration or rapidity of a pendulum increases with the diminution of its length, and is arrested proportionately with the elongation of the rod. Later it was ascertained that such a retardation happens also when the clock is carried on a high mountain. As the vibration of the pendulum is based on the laws of falling bodies, and the fall of the bodies itself depends on their weight or otherwise, on the attraction of the earth (?) it was but natural to conclude that if the vibration of the pendulum not the same everywhere, and the attraction of the earth varies, then this affords us conclusive evidence that the cause of the retardation of the vibrations of the pendulum is certain centrifugal force, which develops with the motion the earth around its axis, and that it is this force, which, arrests the swing of the pendulum by decreasing its weight. But such a conclusion is erroneous and we could far better admit the following conclusion, at which many of our physicists now have arrived — the attraction of the earth diminishes with the recession of the body from its centre, which serves at the same time as the centre for all the attractive force of the globe.

And what if the cause of the retardation of the vibrations of the pendulum at the equator and on high mountain should prove quite different from what is now generally supposed? What if the cause is not at all the decrease of the force of attraction (whether from the recession of the object from the centre of the earth or centrifugal force), but on the contrary, its increase, proceeding from the accumulation of bulk at the equator, in which case the force of attraction increasing, increases at the same time the weight of the body, and in the pendulum the weight of the ball? There is one fact not known to all physicists, I believe, namely, that the rapidity of the vibrations of a pendulum depends not only on the length of its rod, but also on the weight of the ball itself. It might be even more correct to express it thus; the velocity of the motion of the pendulum depends chiefly on the weight of its ball. When I elongate the rod of the pendulum I force the ball to move on a longer level, and increase thereby its own weight; I can also,, without elongating the rod, increase its weight by other means; the result will be the same. Thus, for instance, everyone is aware that even people unacquainted with science, when their clocks are running too fast, and they wish to make the pendulum vibrate slower, attach to the ball either a stone or a small bit of iron, and thus attain their object. The physicists have made very exact experiments in this direction. They found that a pendulum having a uniform length of rod makes 20,000 vibrations—

With a ball attached to it	weighing	2	k.g.	in	1,977	seconds.
,, ,,	,,	4	,,	,,	2,010·55	,,
, ,,	,,	6	,,	,,	2,021·31	,,
,,	,,	8	,,	,,	2,027·04	,,

Therefore the greater weight of the ball the slower vibration of the pendulum. From these experiments, conducted with the greatest precautions and published in the "*Comptes Rendus de l'Academie Francaise,*" tome xxi., p.p. 117-134, it appears: 1. That the laws of Galileo are not quite exact as to the vibrations of the pendulum; 2. That the explanation of the retardation of the pendulum on the equator by the decrease of the force of attraction of the 'earth is evidently false; 3. That even the universally accepted laws of the gravitation of bodies are not sufficiently exact; and 4. That, in general, the means employed toward discovering the laws of nature with the help of calculations is not only being proved unreliable, but it serves but the more to darken the truth.

You will have seen from the last two arguments, which have hitherto served as evidence of the rotation of the earth, that as the result of such a rotation was assumed a centrifugal force. Its presence was vainly sought for in the currents of the ocean, as well as in those of the air. And, indeed, it is not easy to explain how or on what principle the air — this soft, yielding incompressible body, agitated by various currents — could have remained unaffected by the rotation of the terrestrial globe. If the greatest physicists admit that hard bodies are influenced by such a rotation, then it appears, it will not be too bold on my part to maintain that the rotation of the earth around its axis should inevitably exert an influence on the air. This influence should be shown first of all in that, during the rotation of the earth from west to east, there would appear immediately an atmospheric current from east to west.

Indeed, if the earth, together with its atmosphere, rotates in a completely empty space, then in every case it light be possible to admit that the earth rotates without introducing any influence on the atmospheric ocean. But against the theory of such a vacuum we have the very .Quality of the air.

The air, as much as we know of it, has such a great tendency toward expansion that all the hitherto worked out laws of gravitation have remained foreign to it. Were the most exterior, the most rarefied layer of air not to encounter on its way any obstacle toward its expansion in the shape of a new planet, it would scatter itself throughout the whole universe, moving farther and farther into the infinite space; the particles of the air nearer to this layer would follow its example, and, finally, the seas and rivers of the terrestrial globe, all the water would take part in such a process of expansion, to disappear at last from the face of the earth. (We produce first just such a phenomenon with the help of an air pump). On the ground that such a thing does not exist in fact, we must suppose that there is some retaining cause, which according to custom, we will term Ether. Counteraction to the evaporation of the air consists in this, that it forces every upper layer to press upon the next lower, causing by such a progressive pressure the condensation that layer of the atmospheric air which is next to us.

If such an ether exists in reality, then there must occur in the atmosphere those phenomena so familiar to us, which always take place in cases when the air encounters obstacles to its free motion. Let the earth rotate, then all the atmospheric space, on the ground of the attraction of the earth, will be compelled to participate in the movement, and the consequence will be that the upper layers of the air, finding a resistance in the ether, will either be retarded, or — which would be the same — assume a seeming current in a direction opposite to that of the earth's motion. Such a current of the upper stratum of the air would provoke a resistance the next lower one, and this one, in its turn, receiving the impulse communicated to it by the upper one, would offer a resistance to its next lower neighbour, etc. Finally these two opposite currents, intermingling in their onward impulse, would form two streams — one from east to west, in which would participate, first, the whole atmospheric ocean world, and then the contents of all the watery basins; the other from west to east, into which would be drawn the very core of the terrestrial globe.

But let us make another supposition, and notwithstanding the impossibility, let us admit that there is *no* ether; that ether is no more nor less than the product of those endless hypotheses in which man has entangled himself from the first in his efforts to investigate nature; even in the latter case it will not be a difficult task to prove that the rotation of the earth must cause the current of the atmosphere to take an opposite direction, On what ground did our physicists base their suppositions when telling us that we don't feel the rotation of the earth? How do they explain the circumstance that objects on its surface are neither upset nor fall? They point to the laws of inertia. Very well! I agree with them! I agree only the better to vanquish my adversaries with their own weapons, as I have hitherto always done. You are probably aware that motion can be imparted to any substance, but that a fluid or gaseous body can be made to move only when it is imprisoned in a hard one. Air is a body which is more than any other disassociated as to its component parts. Let us suppose that the Earth has communicated its movement to the layer of air next to the surface, and thus dragged it after her. This layer, perfectly separate and distinct from the next upper one unattached to it, is unable to communicate its motion to the other and upper layers. Hence these upper layers remain unaffected by the motion of the lower one, or what comes to the same, begin to assume a *seeming* rush (or current) from east to west, with a rapidity equal to the earth's rotation. Every point of the equator during the diurnal rotation of the earth crosses in the same lapse of time 1,250 feet, but in the direction opposite to that of the earth's rotation. But such a rapidity of the atmospheric currents is nowhere to be seen, and it exceeds ten times the speed of the most terrible hurricanes.

I do not belong to those who accept their own conviction of an east and west atmospheric current for a real and already demonstrated fact. And yet all the modern physicists, scientifically convinced of the absolute necessity for the existence of such a current, have accepted it as a fact, resulting from the earth's rotation around its axis, although all their efforts to find it anywhere in nature have been in vain. Even the *passates,* explained for a certain time by the same rotatory motion of our globe, deprived at the present moment of their once famous periodicity, are now being accounted for a great deal more simply, to wit, by the different degree of heat in the upper envelope of the terrestrial globe.

We have but to represent to ourselves, in thought, all the various atmospheric currents, at one time weakening, at another increasing, and moving in every imaginable direction, called by us sometimes winds, sometimes tempests we must imagine these winds running very often in direct opposition to each other's course, and then ask ourselves the question: Is there any possibility that such currents could exist when the air is at the same time forced to passively follow the simultaneous rotation of the earth around the sun and its own axis? Is it possible to admit that in case such currents existed in nature, our atmosphere would at the same time continue the constant and faithful satellite of our earth?

Therefore the circumstance that the rotation of the earth around its axis is not at all felt by us; that other dream stance, that this rotation has never been in any form or manner satisfactorily proved, and *cannot* be proved; the absence, finally, in nature, of those atmospheric currents which in all justice ought to be found as a consequence of the rotation — all this serves us as a refutation of the theory of the rotation of the earth around its axis, perfectly convincing, if it were only because we do not possess a single evident proof in favour of the rotation.

Is it not a cause of wonder that the *savants* of the whole civilised world, beginning with Copernicus and ending with Kepler, first of all accept such a rotation of our planet, and then for three centuries and a half after that seek for it some proof? But, alas! they seek, and as was to be expected, find it not. All in vain; all unsuccessful!

To prove the impossibility of the second proposition, *i.e.* the revolution of the earth around the sun, will present no difficulty. We can bring self-evident proof to the contrary. *The earth revolves around the sun and is retained in its orbit by the strength of the solar attraction*, and these propositions contradict, point blank, the fundamental law of gravitation itself. It is known to everyone that the direction of the weight is perpendicular to the wall, otherwise the grain of dust would fall. In the same way the direction of the weight of our planet must be perpendicular to the sun, as to the centre of its attraction. But such, in fact, is not the case at all. The direction of the earth's weight is not only not perpendicular, but even changes with every moment.

In order to prove the correctness of my observation, we will now examine more carefully the modern theory of the annual rotation of the earth around the sun, and we will examine it under the aspect in which it is treated in the scientific works that discuss this subject. To explain the change of seasons, in other words to demonstrate the solar ecliptic, the scientists have *assumed* the following position: The earth's axis inclines to its orbit at an angle of 66½° this angle is preserved by the earth during the whole time of its rotation around the sun, *i.e.,* the axis of the earth is parallel to itself at every point of its transit. We can make this theory approximately clear to ourselves by the following illustration: Taking this candle for the sun, we will now revolve around it this little globe, so that, by a simple practical experiment, we may form for ourselves an idea how the four seasons take place... (Diagram I omitted) Here on the diagram we can plainly see that the axis of the earth does not change its position with relation to the earth's orbit during the whole time of the earth's rotation, *i.e.,* it remains parallel to itself. It is only by conceding this that we can explain the four seasons of the year. To this point the modern theory appears perfectly satisfactory, but if we examine it more carefully, its inconsistency will become evident. Thus I will now touch at once that incomprehensible and, at the first glance, unobserved circumstance, which has always appeared to me absurd, whenever I had to explain to my audience the rotation of the earth around the sun.

As it would be absurd to suppose that the sun, during the yearly revolution of the earth, in its turn daily circumscribes the earth, modern theory, to meet the necessity of the case, has to suppose that the terrestrial globe, while rotating yearly around the sun, turns daily around its own axis in the direction from west to east. But such two simultaneous rotations are, as we shall directly see, perfectly inadmissible. During the interval from the 21st of June to the 22nd of September such two simultaneous motions coincide well enough, but from the 22nd of September onward, and back to the 21st of June, the juxtaposition of such two motions carries us on directly to a perfect absurdity; it would follow that the terrestrial globe, rotating diurnally around its axis from west to east, moves onward in a direction quite the opposite. But I believe that everyone is aware that a moving body, according to the nature of its rotary motion, either receives an impulse forward, or, on the contrary, the impulse forward directs its rotary motion. Consequently, if the terrestrial globe rotates from west to east, then it must also proceed onward in the same direction, and, in case of a sudden appearance of some new force, compel the earth to deviate from its primal direction, the force which makes the earth to move around its axis must (if it is the stronger) either overcome the newly manifested force or be destroyed by it.

FIG. II.

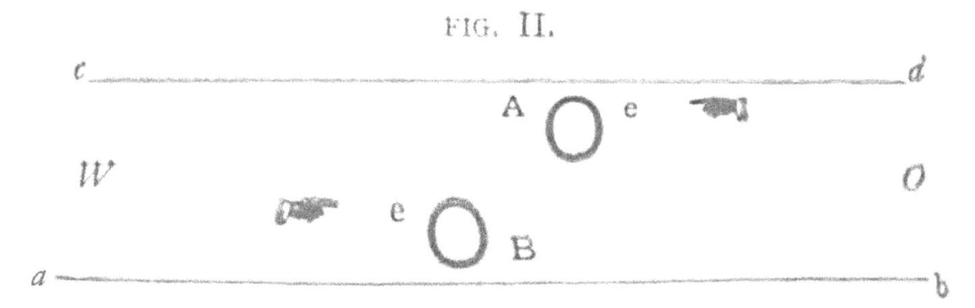

If we compare the two halves (or parts) of the terrestrial revolution around the sun. to wit, the semi-revolution from W to O, through B, with the semi-revolution from O to W, through A, we find that, from W to O, the direction of the rotation agrees to a certain point with the direction of the motion, and from O to W it is directly opposite to its onward motion. This will best be seen if we rotate this sphere around the lighted candle in the same manner as represented for the earth as Fig. I. In order to explain such a strange contradiction we ought to suppose that, during the revolution of the earth around the sun, the direction of the terrestrial weight is also changed, but this would amount to an absurdity, and something in direct contradiction to the accepted formula, that the direction of the terrestrial weight depends on the sun, as on a body which keeps the earth in its orbit. Fig. 2 will explain the whole still plainer. If the globe, e, is compelled to rotate towards O, in the direction pointed to by the hand, and move onward from *a* to *b*, and from *d* to *c*, then, in its motion from W to O, it must have the direction of its weight on the line *a, b,* and in its motion from O to W, on the line *c, d,* to wit, in the first case, have its weight directed downward, and in the second case upward. Although in the universal space there exists neither an up nor down, the question itself is unaffected by that circumstance. Presently we will return once more to this question, and prove that such an incessant change of the direction of the terrestrial weight is in direct contradiction with science.

According to the now prevailing modern view, the earth is kept within its orbit by the force of the sun's attraction. But even this proposition contradicts the assumption of the dual rotation of the earth, unless we make such allowances as will contradict all our scientific notions, for it is impossible to imagine to ourselves two simultaneous motions of the terrestrial globe around its axis, and around the sun, in agreement with the change of years and that of the seasons, during which the direction of the terrestrial weight would be constantly turned toward the sun, as we ought to find it were the earth supported in its orbit by the force of the attraction of the sun. It is supposed that in every circuitous motion there are two forces in action. For instance, if we attach a ball to a string and swing it around so that the cord will be extended out straight, then the one force, which tends to project the ball in a straight line from the centre, is named centrifugal force, and the other, contained in the very cord, shows a tendency to draw back the ball toward the centre round which it revolves, and is called centripetal force. During the simultaneous activity of both the forces the ball cannot move on a direct line on which both forces tend to move it, but is forced to adopt a movement in the direction of a diagonal, and from the union of an infinite number of such diagonals, it begins moving in a circle.

If we examine a little more carefully this circuit-motion of the ball, we will find it anything but complex. That point of the ball to which is attached the cord, *i.e.* near which acts the centripetal force developed by my hand, lies on that side of the ball which is directed to the centre of the movement, *i.e.,* in the direction of the hand, and, if the ball had a propensity at the same time to assume a motion around its axis, then the latter would find itself at the same spot where the thread is tied, and this given point on the ball ought to remain turned toward the hand. That which is law for one body is law for all other bodies, placed in the same conditions as the first. The moon — the only heavenly body so close to our planet as that we can observe it in detail — is placed, in relation to her revolution around the earth, under precisely the same conditions as the ball we are now examining is, in relation to the point where the thread is fixed. Let us fancy the ball as the moon, the hand as the earth, and the thread as the terrestrial attraction, invisible in reality, but acting like the thread, and we will see that the moon is turned toward our globe always on the same side, for the force of attraction has deprived it forever of the slightest possibility to effect any change in the direction of the weight and rotation around its axis. Why, then, not derive from the laws of motion regulating the moon, a very close deduction for our own planet? Indeed if the terrestrial globe revolves around the sun, and is kept in suspension in its orbit through the attraction of the sun, then this globe, as well as the moon, must find it impossible to rotate around its axis. In such a case, the one side of the earth would be constantly lighted by the sun, while the other would find itself in perpetual darkness. But we see no such thing, therefore we must infer that the modern explanations of the movements of our planet around its axis and the sun are devoid of the least probability, and disagree entirely with the exigencies of experiment.

Perhaps we might suppose that the terrestrial globe occupying a. central position, revolves in twenty-four hours around its axis, while the sun describes annually above it that circle which is shown by the ecliptic. But there is no room for such a supposition until the rotation of the earth itself around its axis is demonstrated on more solid proofs; and, besides, as I have shown, it is the contrary, which can be most easily proved. The immobility of our planet is chiefly maintained by me on the principle that we cannot find in Nature any constant atmospheric current always running from east to west. On the same principle, if our planet revolved around the sun, its whole atmosphere ought to be retarded and forced in a direction contrary to the forward motion of the earth, and would have to follow our planet like a long tail, as we see in the case of comets. Of whatever substance may be the tail of the latter, we are forced to examine it as the atmosphere of these as yet but little known bodies, and if the comets themselves travel in the universal space, then their atmosphere is compelled to follow them in the shape of a luminous tail.

Finally, let us return once more to the law of gravitation in order to demonstrate conclusively that the rotation of the earth around its axis and the sun is an utterly improbable hypothesis. A little further back, while repeating to you in substance the theory now thoroughly accepted of the earth's revolution, I have shewn that, as the theory now stands, the position of the terrestrial weight must inevitably be shifting at every second. Out of this would result the following: If the sun really retains the terrestrial globe in its orbit, then the direction of the terrestrial gravity must constantly tend from the centre of the earth toward the point fixed on its surface at that side which is turned to the sun; on this point acts, immediately, all the centripetal force proceeding from the sun, and, therefore, as in the instance of the moon when the centre of all the lunar gravity is concentrated on that side of her which is turned to us, it is to this point that must gravitate all the weight of the terrestrial globe as all the weaker and lighter bodies. But our experiments show to us quite the contrary: the centre of the earth's gravity does not change in the least, and placed in its middle, depends only on the terrestrial mass; no outward force of the kind of the sun's attraction is able to affect it any way, or can force it to displace itself. And if so, then do not such facts prove fully and clearly (1) that the terrestrial globe is not kept in its orbit by the sun's attraction, because such an enormous force could not but affect the point where is concentrated the centre of the earth's gravity; and (2) that the centre of earth is at the same time the centre of its weight, and also the centre of all the visible universe? Of course, I do not reject entirely the influence on our planet not only of the attraction of the sun, but also of the moon, but I only maintain that the force of their attraction is not so powerful as to influence, in any serious way, the solid portions of terrestrial body, when we find that even with fluid and gaseous bodies, especially such as the air, this influence is felt but to a very feeble extent. If the attraction of the sun is so trifling that it can act but in quite a slight and to us as yet not quite clear manner on fluidic bodies, then we have still less reason to suppose that such a weak force could neutralize the centrifugal force of the earth and keep it in its orbit. For such an effect as this a force of gigantic proportion would be required — a force under whose action all the terrestrial atmosphere would long since have been carried off to the sun, in the same way as the force of attraction of the terrestrial globe is ever ready to attract to itself every just-forming lunar atmosphere.

Let us now see what changes would be called for in the same department of astronomy were my assertions to be some day verified, and it should be found that the earth is motionless, and occupies the central position of the visible universe. Such changes would be in some respects important, in others unimportant. They would chiefly consist in our henceforth regarding the hitherto seeming motion of the heavenly bodies as a real motion, as the astronomer Tycho de Brahe did before. He maintained that the earth stands still in the centre of the universe, and around it, as around its natural centre, moves diurnally the whole heavenly sphere; the moon and the sun in addition to the above motion describing around the earth independent movements on special curves, while Mercury with the rest of the planets describes an epicycloid...I may also add that the position assumed by our scientists who consider the fixed stars as suns of the same nature as our own, and all the other planets as bodies identical in substance with our earth, will be found to be without foundation. Such a theory is irrational, if it were only because of the principles on which are based the determination of circumferences and weights of the celestial bodies. The weight of the sun, for instance, was determined in accordance with the amount of the expression of its imaginary attractive force on the surrounding planets. As soon as it is found that the sun must surrender its office of principal star and become simply a planet revolving around the earth, directly depending on the force of the latter's attraction, a will naturally be proved erroneous. The sizes of the heavenly bodies have been determined on no less false principle.

Who but is more or less acquainted with that phenomenon which shows us an object diminishing in proportion to the distance, so that if an object is placed at a distance which exceeds 5,000 times its diameter, the human eye is unable to see that object? It is on the basis of this law that the sizes of all the heavenly bodies have been calculated. According to their seeming size and the ratio of their distance from the earth, science has endeavoured to determine the number of times that their real size surpasses their seeming one. But in determining by that principle our scientists have neglected to consider one of the most important points; they forget that the law which makes objects apparently diminishing in proportion to their distance from the observer does not affect luminous bodies; the brighter the light of the body the longer its bulk will remain unchanged in our sight, whereas an object but faintly lighted becomes invisible, as I have said, at a distance which exceeds its diameter 5,000 times. If the said law extended to luminous bodies, then a flame one inch wide could not be seen at the distance of 225 yards, whereas we know from experiment that the size of its apparent bulk does not change even when the candle is carried to a distance of several thousand yards. As the sunlight is extremely bright, the bulk of the sun must therefore seem unchangeable at an extremely long distance, and IT IS VERY POSSIBLE THAT THE SUN IN REALITY IS BUT LITTLE BIGGER THAN IT SEEMS TO US AT THE DISTANCE. Besides that, it is not only possible but a great deal more plausible to accept the assumption that the laws which shew to us an object diminishing with the distance are applicable only to our own dense atmosphere which surrounds us, and are not operative in a medium so rare as that of the upper spheres. When, after a clear and cold night, the vapours of the air are drawn down to the earth, and the rising sun illuminates the air cleared from the mist, then the mountains, the villages, the environs and edifices, at other times hardly delineated in the blueish atmosphere, suddenly rise before our eyes as if growing up by enchantment; they seem nearer and allow us to examine the slightest details of their structure. In this case the law of the diminution of objects is evidently changed. And there in the ether, in that attenuated matter — or rather let us only speak of ether as empty space — in this vacuum of the universe how can these laws be ever applied?

Generally speaking, as far as I know from personal experience, the *science of optics* is not quite accurate, the sight of the human eye is more or less influenced by the purity of the atmospheric air...

Equally erroneous will be found all the determinations of distances of the fixed stars, once that we have to regard the earth as fixed. According to the now accepted and "wholly dominant theory, on the 21st of December the earth is 40,000,000 miles (185,000,000?) from the point at which it stood on the a 1st of July (June?). On these same dates, with the help of the telescope, directed to one and the same point of the heavens, is observed a certain star which crosses the meridian in the same direction and in the same point of the heavens. It results then that a distance of 40,000,000 miles (185,000,000?) counts as nothing in our comparison of the distance of the observed star! But even such an evident proof of the recision of the fixed stars from the earth loses certainly all its weight if we assume the earth to be motionless.

And now, gentlemen, allow me to lay before you one more contradiction, which, had it been insisted upon before, might have shewn to our scientists long ago the erroneousness of our astronomical calculation. It was found from the determination of the sun's attraction that every body which exerts on the terrestrial globe a pressure of one pound exerts on the sun a pressure of 27 pounds. If all bodies act on the sun with such an increased pressure, it would then seem that the mass of the sun ought to be likewise and in the same proportion more compact than the terrestrial mass, *i.e.,* it would consist of a more dense matter; and yet, comparing the calculations of the weight with those of the circumference of the sun, it has been found that the sun matter is just four times less in density than the substance out of which the earth is formed. The result, then, would be that one and the same body would weigh on the sun 27 times more than when on earth, and its weight would act on the sun 108 times more than it would on our planet; and yet the substance of the sun would present but ¼ of a part of the density of the matter of the terrestrial globe! This, I must say, is incomprehensible to me, and I view such a theory as the result of correct calculations based on a false principle.

I also deny the existence of the atmosphere on any planet whatever. A heavenly body crossing the universe with a velocity hardly comprehensible cannot be possessed of an atmosphere similar to the air of our earth. And here, as before, the moon — a planet with the qualities with which we are best acquainted — gives us a fully correct comprehension, or rather it corroborates all that is shown to us by the natural laws. The moon has no atmosphere, and, therefore, there is but little probability that the other planets would have any more than she has. All the observations tending to show that the moon must have an atmosphere are based, no doubt, on equally erroneous principles; they could be accepted with any degree of certainty only when the experimenter could be carried beyond the atmosphere of the earth, or, at the least, when we should build our observations on the summit of Dhawalaghiri. The outer services of the body of the sun, moon and other planets cannot be similar in appearance to the surface of the terrestrial globe; they must consist of strongly compacted matter, such as we see sometimes in the substance of the frequently falling aerolites. All the non-solid bodies, the strata of the earth, and the rocky portions would be torn off and precipitated on the earth by the force of its attraction. Thus, on the ground of these premises, the assumption that some of the planets may be inhabited is void of any probability and has to pass into the realm of fiction...

Man, while determining the distance of the stars most important to us, on the strength of an imaginary rule of distance and falsely applied laws of the diminution of objects in proportion to their recession, began to calculate the size of these stars, and, astonished at their dimensions, mistook the fixed stars for bodies similar to our sun, and our earth for a very unimportant portion of the whole universe. Arrived at the latter conclusion, it very naturally appeared absurd to him that all these powerful, all these gigantic and numerous celestial bodies should revolve around our little globe, obey it, and submit to its desires. At that time appeared a new hypothesis: the earth is not motionless, it revolves around itself and around the sun. This theory is accepted as the correct one, and step after step are now built new suppositions, new combinations deduced from the union and combination of imagination with correct mathematical calculations.

Here I end my dissertation, although it would be but an easy matter to point out a great many more contradictions on which rests the modern theory which I now combat and is opposed to mine, We cannot help desiring and hoping that perchance there may be found at least one astronomer who, armed with all the weapons of modern speculative science and its apparatus, will undertake to recreate the whole system of Tycho de Brahe. The result of such an attempt would doubtless prove something scientifically grand. All that now under the Copernican system appears to us so incomprehensible and diametrically opposed to the fundamental laws of nature would be finally explained in the simplest and most rational way. We can now see how right was the venerated astronomer Bandes, when expressing his opinion on Tycho de Brahe's system, he remarked: "This theory presents in itself a great deal more of probability, as it explains so well all of the individual phenomena of nature." Unfortunately, Bandes was mistaken when he imagined that this system contradicted the laws of attraction. But I believe I have fully disposed of such a misunderstanding, and proved that it was not Tycho de Brahe's system, but that of Copernicus, which contradicts all the laws of gravitation.

To add a few more proofs to our assumption we will say:

1. That the form of the continents contradicts the theory of the rotation of the earth. If our globe were revolving around its axis, then the outlines of the continents ought to elongate themselves in a direction from east to west, when in reality this elongation of configuration extends from north to south.

Besides that, the width of their northern edges arises from the attractive force of the northern pole, and the points turned south from the repulsive force of the south pole.

2. There are no fixed stars in the sense of this word, because it has been observed that these stars, besides their diurnal revolution around the earth, perform independent circuitous movements. Vain have been all the efforts of the astronomers to find a central body whose force of attraction might account for the fact that these stars are kept within their orbits; and such a body must exist somewhere. This central body is our earth. May it not also explain the fact that the greater the accumulation of soil in the northern hemisphere the larger is the number of stars above?

3. Various changes in the fixed stars have been often remarked, namely a change of colour or the intensity of light, and sudden appearance and as sudden disappearance of single stars — which does not at all agree with the assumption that they are as large and independent bodies as it has been hitherto supposed.

4. The similarity in the component parts of all the meteorological masses, that is to say, of the bodies attracted by the force of gravity within the earth's atmosphere, gives us chiefly some idea of composition of the mass of all the heavenly bodies, and proves that they cannot be inhabited. The greatest aerolites known to us had a diameter of 7 to 7½ feet.

5. According to the exact researches of Wilhelm-Malman, in the middle latitudes of the temperate zone the prevailing atmospheric current appears to be W.S.W. Although agreeably with the law of terrestrial rotation the prevailing winds ought to be found in those regions easterly, we see the contrary and find them westerly.

As my following work will tend to demonstrate this agreement in the progression of the creation of the universe with truth and fact, and taking into consideration that this pamphlet of mine (the only reasonable refutation of the earth's rotation) shows a similarity with the opinions of many scientists who preceded me, in conclusion I wish to quote a few words from Goethe. The poet, whose prophetic views remained during his life wholly unnoticed, said the following: "In whatever way or manner may have occurred this business, I must still say that I curse this modern theory of cosmogony, and hope that perchance there may appear in due time some young scientist of genius who will pick up courage enough to upset this universally disseminated delirium of lunatics." ...*From the "Scientific American" April 27th, 1878.*

A Vindication of the Divine Cosmogony

By John Dove, M.A. (1757).

That Moses was acquainted with the most abysmal mysteries of Nature is a truth denied by none but upstart philosophers, who would revile him without having read or understood him.

The three first chapters of Genesis contain a revelation of what otherwise would never have been known, *i.e.,* the first principles or rudiments of knowledge, natural and divine. But for the information recorded in those chapters, the human race had never known science or anything concerning the facts of creation. For we were created; there is nothing innate in us or derived from prior existences; language itself was given, not acquired. The philosopher who pleads for any other cause than a divine creation, simply writes himself down a fool. It is useless for the genuine truth-seeker to expect to derive information from those who will need write before they have read; or from the commentators who will give every sense of the text but the true one; or from the system-mongers who will cripple the whole Scripture to make it speak their sense; nor from the philosophers who believe they know better than the inspired historians, or argue that there is no certain standard of truth and that we were sent hither to grope in the dark or learn wisdom from our fellow worms, Moses affirms: "In the beginning God made the heavens and the earth"; the philosophers maintain the eternity of matter, make a god of it, and bow down to the idol they have set up, and would, like Nebuchadnezzar, put everyone in a furnace who refuses obedience to their decrees! To listen to their description of gravity, attraction, centrifugal and centripetal forces, it would carry the appearance of a romance. Did any man yet ever understand Sir Isaac Newton's philosophy; or will any man undertake to prove the truth of it? His warmest advocates have acknowledged "they had not all that evidence of its truth that they could desire"; because they have rejected the revelation of God, and have set up they know not what. They are incorrigible and will not be corrected. Therefore I quit them all and turn to the ecclesiastics, whose proper business it is to study and expound the Scriptures, But I have to tell them as well as the philosophers that in rejecting or doubting the book of Genesis, they stumble at the very threshold of their studies, and seldom or ever after recover themselves. If they understood or believed in Moses, they would possess more real knowledge than all their other learning can teach them.

It is or should be a matter granted, that God and His works must agree; therefore, he that fully understands any part of God's works of creation, as seen in the visible world, and can find in the account given of them in Moses, the Prophets or the Apostles any disagreement, has a right, as a rational creature to be a Deist; but if no such disagreement can be found, instead of a rational Deist, he must be a fool. And since it is truth, that philosophy and divinity are closely connected, and that an error in the former cannot, in producing an error in the latter; and since no system was philosophy, in any age, hitherto proposed to mankind, besides that of Moses, was ever pretended to agree with Scripture, — it is not very extraordinary that no philosopher who pretended to have any respect for the Scriptures, has ever attempted to understand and compare the philosophy of Moses with the real and demonstrable facts of nature? Can it be for want of ability, or that they wilfully prefer falsehood to truth, in the hope or belief that others would do the same? If what Moses wrote was not the literal truth, why have not his mistakes been honestly pointed out by our gentlemen of science? Moses has given us a rational process of the creation, which is more than any one else has done, and more may be said of him than any other philosopher that ever lived, viz., that he has not made one mistake in the account he has given of nature; all the others have scarce delivered one truth concerning it! Truth and falsehood can never be made to agree; therefore, all the experiments that the modern philosopher can make, will never make their system agree with truth or common sense; but they all demonstrate the truth of the Mosaic account of Nature!

The revelation of God is plain, not delivered in mysterious language, as is the modern philosophy, and, when understood, corresponds with right reason. Is it not therefore strange that so many disagreements of it should still subsist? For I cannot find that men in general know any more about it, than about the laws and language of the world in the moon, if such a world there be.

In the two first chapters of Genesis, Moses has given a distinct and positive statement of the mechanical laws or operations by which nature rose into being by the hands of her omnipotent Creator, and by which her stupendous works are still carried on; for nature came not into being by chance or from any pre-existing condition; nor was any fact stated which is not open to the examination of every intelligent person, but which no man yet, has been able to overthrow or improve upon.

But what a condition are we in at present? Not one dignitary in Europe, that has learning or honesty enough to determine the truth of these divine records! Is it possible to conceive that both Protestants and Papists have agreed to let the people be under such delusions? An absolutely correct and literal translation of the Hebrew Scriptures I would present to our view one uniform system of divine, moral, and philosophical truth, that would dispel error, as the morning dawn scatters the darkness of the night. So, then, as all that truth which the faith of a Christian has anything to do with, is contained in Scriptures of Moses, the Prophets, and Apostles, whatever agrees not with those Scriptures is to be rejected, whether it relates to divinity or philosophy. For if in them we have false accounts of the Works of God, no man in his senses will or ought to believe they contain a revelation of God. What! Shall the God of truth not give us a true account of His own work. Shall the God of Nature deceive our senses? God forbid! For as we can know nothing of God but by His Works, nor of His Works, till they are apprehended by the senses He has given us, it is utterly inconceivable to suppose He should have endowed us with such senses as are only calculated to deceive us, or by giving a false account of the works of His own hand.

If, in the language this revelation was originally made, our opponents can find but one philosophical mistake we will unreservedly yield up the whole for a cheat! The translators and the whole group of commentators are herein to blame; for they have all to a man been blinded by a false philosophy, and have resented every attempt to unshackle them; whereby they have been bewildered in uncertainty and error, and have left their readers in darkness and bondage ever since.

Are there any abettors of this heathen philosophy still amongst us? Yes, ten thousand; not only among the unlearned, but amongst our church dignitaries, our classical scholars and teachers! All on account of their ignorance and unbelief.

What will be the end of these things! I am no conjurer; but it is easy to determine what will be, from what has already taken place. It has been the fate of all kingdoms, nations, and people, from the beginning of time, upon their rejecting or perverting the revelation of God, to fall into anarchy, confusion and infidelity. The Bible is, as it deserves to be, the great charter of our liberty. The loss of the Scriptures, or swerving from, or perverting the doctrines or history contained in them, has invariably been attended with discomfiture and ruin, and always will! And if their successors continue their resistance as they have done hitherto, it cannot fail to deluge the kingdom in atheism, destroying all social virtue, and turning it into a field of blood.

The system the philosophers would establish is founded on quicksand, on a spirit of falsehood and lies; its stones unhewn—its mortar untempered—and its joints all open to the weather; when the winds blow, and the floods of opposition beat against it, it must tumble down and disappoint the faith of those dupes who trusted in its strength; because it is not founded nor erected according to, but against, the appointment and design of the Creator. The Scriptures contain the instructions of God, and show us the conditions, the ordinances, the laws which He hath ordained.

I have to repeat, again and again, that the Scriptures and nature are connected, as will appear to any impartial inquirer; those who will not take the pains to study them both, will remain fools, whether I say so or not. The not attending to this connexion has been the cause of that contempt with which the Scripture has been treated. Suppose we view the dial plate of a watch, we see the hand point to the hour, by a mechanism to us invisible; but we find a book wherein the inward structure of the watch or clock is described; we are at a loss whether to believe it or not; we know not whether it be true or false. How then shall we prove its truth! By taking the machine to pieces, and examining its works; if the book and the machine exactly agree, and the former be an accurate description of the latter, the inference must be, that either the maker of the machine wrote the book, or revealed the mechanism of it to him who did. This is absolutely the case between the Bible and nature. And if this examination were firmly, and candidly, and intelligently carried through, the numbers of our foolish philosophers would soon be diminished, and their specious system utterly confounded. Moses and the Prophets never revealed the proper frame of a mouse-trap or the size of a bird cage, because they knew the star gazers would not heed such trifles, nor find any credit in constructing such things. But Moses and the Prophets did, by the inspiration and dictation of God, reveal to mankind the framework and mechanism of nature, which must have remained for ever inscrutable, but for such direct revelation; and which mode and plan of creation, when thus made known, appears true upon the highest demonstration the rational mind can demand!

Now for a coat of mail, to defend me from the tongues of scorpions, and the quills of porcupines, — a venomous serpentine brood, who besmear and befoul every divine and scriptural truth that runs counter to their almighty decrees. Let any man read those mystical and philosophical expostulations between God and Job; or let him read over both Testaments, and he shall find, if he reads attentively, that Scripture, all the way, makes use of nature, and hath revealed such mysteries as are not to be found in all the philosophers; so that I fear not to say that nature is so much the business of Scripture, that the spirit of God, in those sacred oracles, seems not only to dwell on the restitution of man in particular, but even the redemption of nature in general, and is as jealous of the right understanding of the one as of the other.

To speak then of God, without Nature, is more than we can do, tor he is not known in this way; and to speak of Nature without God, is more than we may do; for we should be robbing God of His glory, and attribute those effects to

Nature, which belong only to God and to His spirit which works in Nature. No man can venture to complain ii we use Scripture to prove philosophy, and philosophy to prove divinity; because there is no divinity without nature, nor any true philosophy without God. It is a union insisted on by God, however objected to by man.

If men would but take Mr. Locke's advice, and have the modesty to settle the limits of their understandings and determine what objects lay beyond, and what within their reach, they would not venture so often at things too high for them; of if they had the humility to consult Moses, he would prevent much fruitless labour and correct much inexcusable ignorance.

Real Christian philosophy is a pure and ennobling study, exalting the mind, and lifting it above every sordid pursuit, above everything that is low, little, or mean.

List of Works Quoted

A

Atlas of Physical Geography.
Age, The
Anson's Voyage round the World,
Astronomy, Six Lectures on Answers.
Argus Annual, The
Astronomy.
American Exporter.
Answers to Planar Questions.
Act of Parliament.
Age of Reason, The
Agnostic Journal, The
Antiquities of the Jews.
Anti-Infidel—App.

B

Bible versus Science.
The Belfast News Letter.
Birmingham Weekly Mercury.

C

Christian Million, The
Ceylon Independent, The
Chambers' Mathematical Tables.
Chambers' Information for the People
Chambers' Journal.
Cowper, The Poet
Cause of an Ice Age, The
Cape Times, The
Cape Argus, The
Cruise of the Falcon, The
Cook's Voyages.
Christian World Pulpit.

D

Daily Chronicle, The
Daily News, The

E

Elementary Physiography
English Mechanic, The
Earth not-a-globe Review,
Errors of Geologists
Engineer of P. W. D., Capetown
Encyclopedia Britannica
Extracts from Rev. J. Wesley's Works.
Echo, The
Earth and its evidences, The

F

Friend, Dr. W.
Figures of the earth.
First principles of Natural Philosophy.
Field, The
Flood and Geology, The
Future, The
Fifty Scientific Facts.
Freethinker, The

G

Geological Journal, The
Geology.
Geologists, Errors of

H

History of the Conflict between Religion and Science.
History of the Gr. Western Railway.
Harper's Weekly.

J

Journal of Society of Arts.
Joshua's long day.
Joshua commanding the Sun to stand still.

K

Knowledge.

L

London Journal, The
Land of the Midnight Sun, The
Lancet, The
Liverpool Mercury, The
Lux.
Leicester Daily Post.
Lucifer.

M

Modern Science and Modern Thought by S. Laing.
Modern Review.
Mensuration by T. Baker, C.E.
Million of Facts, A
Mill, J. S.
More Worlds than one.
Music and Morals.
Museum of Science.
Magnetism and Deviation of the Compass
Magnetism.
Magnetism and Electricity.
Morning Leader.
Muses, The

N

Navigation by D. W. Barker and W. Allingham.
Nature and Man.
Natal Mercury.
New Principia, The
Noad's Lectures on Chemistry.
Nineteenth Century.
Nautical Almanac.
Navigation and Nautical Astronomy.
Navigation in Theory and Practice.
Navigation and Nautical Astronomy, by Bergen.
Navigation, by Evers.
Naval Science.
New-York Independent.

O

Our place among infinities, by R. A. Proctor
Omoo, by H. Melville.

P

Principia, The, by Sir I. Newton.
Proceedings of the Royal Institute, Great Britain.
Popular Science Recreations.
Pagan Astronomy.
Pollock, N. H.
Primer of Navigation, A
Pearson's Weekly,
Paul Petoff.

R

Recollections of Past life.
Reynold's Newspaper.
Robinson's New Navigation and Surveying.

S

Science Siftings
Science and Culture by Prof. Huxley.
Story of the Heavens, by Sir R. Ball.
Standard, The
South Sea Voyages.
Six Lectures on Astronomy.
Sun, Moon and Stars.
Scientific American.
Scraps.
Skellam, A. E.
Sun, The
Story of the Solar System.

T

Travels in the Air, by J. Glaisher.
Triumph of Philosophy, by J. Gillespie.
Theoretical Astronomy.
Treatise on Astronomy, A.

V

Voyage of a Naturalist, by C. Darwin.
Voyage towards the South Pole.

W

Western Christian Advocate, The
Wonders of the Sun, Moon and Stars.

Y

"Yellow Frigate," by J. Grant.

Z

Zetetic Astronomy, by Parallax.

www.ingramcontent.com/pod-product-compliance
Lightning Source LLC
Chambersburg PA
CBHW022106120125
20250CB00031B/503

* 9 7 8 1 7 8 9 8 7 2 9 8 9 *